Research Notes in Mathematics

Main Editors
A. Jeffrey, University of Newcastle-upon-Tyne
R. G. Douglas, State University of New York at Stony Brook

Editorial Board
F. F. Bonsall, University of Edinburgh
H. Brezis, Université de Paris
R. J. Elliott, University of Hull
G. Fichera, Università di Roma
R. P. Gilbert, University of Delaware
K. Kirchgässner, Universität Stuttgart
B. Lawson, State University of New York at Stony Brook
R. E. Meyer, University of Wisconsin-Madison
J. Nitsche, Universität Freiburg
L. E. Payne, Cornell University
G. F. Roach, University of Strathclyde
I. N. Stewart, University of Warwick
S. J. Taylor, University of Virginia

Submission of proposals for consideration
Suggestions for publication, in the form of outlines and representative samples, are invited by the editorial board for assessment. Intending authors should contact either the main editor or another member of the editorial board, citing the relevant AMS subject classifications. Refereeing is by members of the board and other mathematical authorities in the topic concerned, located throughout the world.

Preparation of accepted manuscripts
On acceptance of a proposal, the publisher will supply full instructions for the preparation of manuscripts in a form suitable for direct photo-lithographic reproduction. Specially printed grid sheets are provided and a contribution is offered by the publisher towards the cost of typing.

Illustrations should be prepared by the authors, ready for direct reproduction without further improvement. The use of hand-drawn symbols should be avoided wherever possible, in order to maintain maximum clarity of the text.

The publisher will be pleased to give any guidance necessary during the preparation of a typescript, and will be happy to answer any queries.

Important note
In order to avoid later retyping, intending authors are strongly urged not to begin final preparation of a typescript before receiving the publisher's guidelines and special paper. In this way it is hoped to preserve the uniform appearance of the series.

Titles in this series

1 Improperly posed boundary value problems
A Carasso and A P Stone
2 Lie algebras generated by finite dimensional ideals
I N Stewart
3 Bifurcation problems in nonlinear elasticity
R W Dickey
4 Partial differential equations in the complex domain
D L Colton
5 Quasilinear hyperbolic systems and waves
A Jeffrey
6 Solution of boundary value problems by the method of integral operators
D L Colton
7 Taylor expansions and catastrophes
T Poston and I N Stewart
8 Function theoretic methods in differential equations
R P Gilbert and R J Weinacht
9 Differential topology with a view to applications
D R J Chillingworth
10 Characteristic classes of foliations
H V Pittie
11 Stochastic integration and generalized martingales
A U Kussmaul
12 Zeta-functions: An introduction to algebraic geometry
A D Thomas
13 Explicit *a priori* inequalities with applications to boundary value problems
V G Sigillito
14 Nonlinear diffusion
W E Fitzgibbon III and H F Walker
15 Unsolved problems concerning lattice points
J Hammer
16 Edge-colourings of graphs
S Fiorini and R J Wilson
17 Nonlinear analysis and mechanics: Heriot-Watt Symposium Volume I
R J Knops
18 Actions of fine abelian groups
C Kosniowski
19 Closed graph theorems and webbed spaces
M De Wilde
20 Singular perturbation techniques applied to integro-differential equations
H Grabmüller
21 Retarded functional differential equations: A global point of view
S E A Mohammed
22 Multiparameter spectral theory in Hilbert space
B D Sleeman
24 Mathematical modelling techniques
R Aris
25 Singular points of smooth mappings
C G Gibson
26 Nonlinear evolution equations solvable by the spectral transform
F Calogero
27 Nonlinear analysis and mechanics: Heriot-Watt Symposium Volume II
R J Knops
28 Constructive functional analysis
D S Bridges
29 Elongational flows: Aspects of the behaviour of model elasticoviscous fluids
C J S Petrie
30 Nonlinear analysis and mechanics: Heriot-Watt Symposium Volume III
R J Knops
31 Fractional calculus and integral transforms of generalized functions
A C McBride
32 Complex manifold techniques in theoretical physics
D E Lerner and P D Sommers
33 Hilbert's third problem: scissors congruence
C-H Sah
34 Graph theory and combinatorics
R J Wilson
35 The Tricomi equation with applications to the theory of plane transonic flow
A R Manwell
36 Abstract differential equations
S D Zaidman
37 Advances in twistor theory
L P Hughston and R S Ward
38 Operator theory and functional analysis
I Erdelyi
39 Nonlinear analysis and mechanics: Heriot-Watt Symposium Volume IV
R J Knops
40 Singular systems of differential equations
S L Campbell
41 N-dimensional crystallography
R L E Schwarzenberger
42 Nonlinear partial differential equations in physical problems
D Graffi
43 Shifts and periodicity for right invertible operators
D Przeworska-Rolewicz
44 Rings with chain conditions
A W Chatters and C R Hajarnavis
45 Moduli, deformations and classifications of compact complex manifolds
D Sundararaman
46 Nonlinear problems of analysis in geometry and mechanics
M Atteia, D Bancel and I Gumowski
47 Algorithmic methods in optimal control
W A Gruver and E Sachs
48 Abstract Cauchy problems and functional differential equations
F Kappel and W Schappacher
49 Sequence spaces
W H Ruckle
50 Recent contributions to nonlinear partial differential equations
H Berestycki and H Brezis
51 Subnormal operators
J B Conway
52 Wave propagation in viscoelastic media
F Mainardi
53 Nonlinear partial differential equations and their applications: Collège de France Seminar. Volume I
H Brezis and J L Lions
54 Geometry of Coxeter groups
H Hiller
55 Cusps of Gauss mappings
T Banchoff, T Gaffney and C McCrory

56 An approach to algebraic K-theory
A J Berrick
57 Convex analysis and optimization
J-P Aubin and R B Vintner
58 Convex analysis with applications in the differentiation of convex functions
J R Giles
59 Weak and variational methods for moving boundary problems
C M Elliott and J R Ockendon
60 Nonlinear partial differential equations and their applications: Collège de France Seminar. Volume II
H Brezis and J L Lions
61 Singular systems of differential equations II
S L Campbell
62 Rates of convergence in the central limit theorem
Peter Hall
63 Solution of differential equations by means of one-parameter groups
J M Hill
64 Hankel operators on Hilbert space
S C Power
65 Schrödinger-type operators with continuous spectra
M S P Eastham and H Kalf
66 Recent applications of generalized inverses
S L Campbell
67 Riesz and Fredholm theory in Banach algebra
B A Barnes, G J Murphy, M R F Smyth and T T West
68 Evolution equations and their applications
F Kappel and W Schappacher
69 Generalized solutions of Hamilton-Jacobi equations
P L Lions
70 Nonlinear partial differential equations and their applications: Collège de France Seminar. Volume III
H Brezis and J L Lions
71 Spectral theory and wave operators for the Schrödinger equation
A M Berthier
72 Approximation of Hilbert space operators I
D A Herrero
73 Vector valued Nevanlinna Theory
H J W Ziegler
74 Instability, nonexistence and weighted energy methods in fluid dynamics and related theories
B Straughan
75 Local bifurcation and symmetry
A Vanderbauwhede
76 Clifford analysis
F Brackx, R Delanghe and F Sommen
77 Nonlinear equivalence, reduction of PDEs to ODEs and fast convergent numerical methods
E E Rosinger
78 Free boundary problems, theory and applications. Volume I
A Fasano and M Primicerio
79 Free boundary problems, theory and applications. Volume II
A Fasano and M Primicerio
80 Symplectic geometry
A Crumeyrolle and J Grifone
81 An algorithmic analysis of a communication model with retransmission of flawed messages
D M Lucantoni
82 Geometric games and their applications
W H Ruckle
83 Additive groups of rings
S Feigelstock
84 Nonlinear partial differential equations and their applications: Collège de France Seminar. Volume IV
H Brezis and J L Lions
85 Multiplicative functionals on topological algebras
T Husain
86 Hamilton-Jacobi equations in Hilbert spaces
V Barbu and G Da Prato
87 Harmonic maps with symmetry, harmonic morphisms and deformations of metrics
P Baird
88 Similarity solutions of nonlinear partial differential equations
L Dresner
89 Contributions to nonlinear partial differential equations
C Bardos, A Damlamian, J I Díaz and J Hernández
90 Banach and Hilbert spaces of vector-valued functions
J Burbea and P Masani
91 Control and observation of neutral systems
D Salamon
92 Banach bundles, Banach modules and automorphisms of C*-algebras
M J Dupré and R M Gillette
93 Nonlinear partial differential equations and their applications: Collège de France Seminar. Volume V
H Brezis and J L Lions
94 Computer algebra in applied mathematics: an introduction to MACSYMA
R H Rand
95 Advances in nonlinear waves. Volume I
L Debnath
96 FC-groups
M J Tomkinson
97 Topics in relaxation and ellipsoidal methods
M Akgül
98 Analogue of the group algebra for topological semigroups
H Dzinotyiweyi
99 Stochastic functional differential equations
S E A Mohammed
100 Optimal control of variational inequalities
V Barbu
101 Partial differential equations and dynamical systems
W E Fitzgibbon III
102 Approximation of Hilbert space operators. Volume II
C Apostol, L A Fialkow, D A Herrero and D Voiculescu
103 Nondiscrete induction and iterative processes
V Ptak and F-A Potra
104 Analytic functions – growth aspects
O P Juneja and G P Kapoor
105 Theory of Tikhonov regularization for Fredholm equations of the first kind
C W Groetsch

106 Nonlinear partial differential equations and free boundaries
J I Díaz
107 Tight and taut immersions of manifolds
T E Cecil and P J Ryan
108 A layering method for viscous, incompressible L_p flows occupying R^n
A Douglis and E B Fabes
109 Nonlinear partial differential equations and their applications: Collège de France Seminar. Volume VI
H Brezis and J L Lions
110 Finite generalized quadrangles
S E Payne and J A Thas
111 Advances in nonlinear waves. Volume II
L Debnath
112 Topics in several complex variables
E Ramírez de Arellano and D Sundararaman
113 Differential equations, flow invariance and applications
N H Pavel
114 Geometrical combinatorics
F C Holroyd and R J Wilson
115 Generators of strongly continuous semigroups
J A van Casteren
116 Growth of algebras and Gelfand–Kirillov dimension
G R Krause and T H Lenagan
117 Theory of bases and cones
P K Kamthan and M Gupta
118 Linear groups and permutations
A R Camina and E A Whelan
119 General Wiener–Hopf factorization methods
F-O Speck
120 Free boundary problems: applications and theory, Volume III
A Bossavit, A Damlamian and M Fremond
121 Free boundary problems: applications and theory, Volume IV
A Bossavit, A Damlamian and M Fremond
122 Nonlinear partial differential equations and their applications: Collège de France Seminar. Volume VII
H Brezis and J L Lions
123 Geometric methods in operator algebras
H Araki and E G Effros
124 Infinite dimensional analysis – stochastic processes
S Albeverio

E Ramírez de Arellano &
D Sundararaman (Editors)
Instituto Politécnico Nacional, Mexico

Topics in several complex variables

Pitman Advanced Publishing Program
BOSTON · LONDON · MELBOURNE

PITMAN PUBLISHING INC
1020 Plain Street, Marshfield, Massachusetts 02050

PITMAN PUBLISHING LIMITED
128 Long Acre, London WC2E 9AN

Associated Companies
Pitman Publishing Pty Ltd, Melbourne
Pitman Publishing New Zealand Ltd, Wellington
Copp Clark Pitman, Toronto

First published 1985

AMS Subject Classifications: 32-06, 53-06

ISSN 0743-0337

Library of Congress Cataloging in Publication Data
Main entry under title:

Topics in several complex variables.

"Pitman advanced publishing program."
"Taller de Varias Variables Complejas (III Coloquio del Departamento de Matemáticas), Centro de Investigación del IPN, Centro Vacacional 'La Trinidad', Santa Cruz Tlaxcala, México, 15 al 19 de agosto de 1983"—
1. Functions of several complex variables—Addresses, essays, lectures. I. Ramírez de Arellano, E. (Enrique) II. Sundararaman, D. III. Instituto Politécnico Nacional (Mexico). Departamento de Matemáticas. IV. Taller de Varias Variables Complejas (1983 : Santa Cruz Tlaxcala, Mexico)
QA331.T5753 1985 515.9′4 84-14858
ISBN 0-273-08656-1

British Library Cataloguing in Publication Data

Topics in several complex variables.
1. Functions of complex variables
I. Ramírez de Arellano, E.
II. Sundararaman, D.
515′.9 QA331

ISBN 0-273-08656-1

Reproduced and printed by photolithography
in Great Britain by Biddles Ltd, Guildford

Contents

COURSES OF LECTURES

M. Kuranishi:
Local Geometry of Nondegenerate CR Structures 1

H. Rossi:
Applications of Complex Analysis to Group Representations........ 37

B. Shiffman and J. Sommese:
Vanishing Theorems for Weakly Positive Vector Bundles 61

S.M. Webster:
Real Submanifolds of $\mathbb{C}^n$ and Their Complexifications........ 69

LECTURES

Al. Boggess:
A Survey of Recent CR Extension Results........ 80

L. Brambila:
Endomorphisms of Vector Bundles Over a Compact Riemann Surface 90

R. Ephraim:
A Generalized Criterion for Equisingularity........ 96

M.A. Guest:
Geodesics, Harmonic Maps, and The Yang-Mills Equations 102

M. Kalka:
Deformations of Submanifolds and Vanishing Theorems........ 109

A. Markoe:
Computerized Tomography and Complex Analysis 113

A. Nagel:
Non-Isotropic Metrics on Boundaries of Domains of Finite Type........ 126

M. Porter:
Properties Related to Directional Convexity 136

J. Ramanathan:
Harmonic Maps from Surfaces to The Complex Grassmann Manifolds........ 148

J. A. Seade:

Vector Fields on Smoothings of Complex Singularities 152

M. J. Spurr:

On the Genus of An Irreducible Component of the Zero Set of A Holomorphic One-Form 158

D. Sundararaman:

Construction of The Normal Cartan Connection (d'aprés M. Kuranishi) 162

L. M. Tovar:

Open Stein Subsets and Domains of Holomorphy in Complex Spaces 183

Preface

The Workshop On Several Complex Variables was held in the Centro Vacacional "La Trinidad", Santa Cruz, Tlaxcala, Mexico, from August 15 to 19, 1983. The Workshop formed part of the activities of the Third Coloquio de Matemáticas held by the Mathematics Department of the Centro de Investigación y de Estudios Avanzados del Instituto Politécnico Nacional, Mexico City.

This volume contains the Proceedings of the Workshop and also the series of lectures given by Professor Masatake Kuranishi in our department in 1982. We wish to thank the participants in the Workshop, the contributors to this volume and the referees.

We express our appreciation and thanks to the Consejo Nacional de Ciancia y Tecnología, the Secretaría de Educación Pública, the Instituto Politécnico Nacional and the Centro de Investigación y de Estudios Avanzados for their support to the Coloquio and the Workshop.

Enrique Ramírez de Arellano Duraiswamy Sundararaman

August, 1984 CINVESTAV-IPN Mexico City

M KURANISHI

Local geometry of nondegenerate CR structures

Our aim in these lectures is to give a detailed construction of the normal Cartan connection for CR structures with non-degenerate Levi-form. The method we adopt is the one given in S. S. Chern and J. K. Moser [1]. These lectures were given during the second semester of 1982 in the Department of Mathematics, Centro de Investigación y Estudios Avanzados del IPN, Mexico City.

§ **1.** In constructing the geometry of CR structures, we exploit the features in the structures analogous to the hermitian geometric structures. Hence we start by pointing out such an analogy. Note that in a hermitian manifold M, we have a complex structure, i.e., a decomposition

$$\mathbb{C}TM = T''\oplus T', \qquad T''\cap T' = \{0\}, \qquad T' = \overline{T''},$$

where T'' satisfies the integrability conditions. In addition we have the non-degenerate hermitian form

$$T''\times T'' \longrightarrow M\times\mathbb{C}.$$

In the case of a non-degenerate CR manifold M, we have

$$\text{(1)} \qquad \text{fiber-dim}\,\mathbb{C}TM/(T'\oplus T'') = 1, \qquad T''\cap T' = \{0\}, \qquad T' = \overline{T''},$$

where T'' satisfies the integrability conditions. The Levi-form is a non-degenerate hermitian form

$$\text{(2)} \qquad L\colon T''\times T'' \longrightarrow \mathbb{C}TM/(T'\oplus T'')$$

In terms of sections, the above map is given by

$$X\times Y \longrightarrow i[X, \overline{Y}] \qquad \text{modulo } T'\oplus T''.$$

Now the analogy is obvious.

A best approach to exploit the above analogy for non-degenerate CR structures seems to be that of E. Cartan, which relies on Cartan connections. For simplicity we recall the approach in the case of Riemann geometry.

Let M be a Riemannian manifold. B be the set of orthonormal coframes; $\rho\colon B \longrightarrow M$ be the source map. B is a principal bundle with structure group $O(n)$.

A chart of B: Pick a smooth section $\xi(x) = (\xi^1(x), \ldots, \xi^n(x))$ of orthonormal coframes. Then

$$B \ni \eta = (\eta^1, ..., \eta^n) \quad \text{at } x \longleftarrow (x, w_j^i),$$

$(w_j^i) \in O(n)$ by $\eta^i = w_j^i \xi^j(x)$.

We define Pfaffian forms $(\omega^1, ..., \omega^n)$:

$$(\omega^i)_\eta(\dot{x}) = \eta^i(dp\dot{x}) \qquad \text{(Tautology forms)}.$$

In terms of the above chart,

$$\omega^i = w_j^i \xi^j(x).$$

Then

$$d\omega^i = dw_j^i \wedge \xi^j + w_j^i d\xi^j$$
$$= \mu_j^i \omega^j,$$

$\mu_j^i \equiv \sum_k dw_k^i \wedge w_k^j, \mod(dx^1, ..., dx^n)$.

However, μ_j^i as above is not unique. In fact, we can consider any

$$\omega_j^i = \mu_j^i - A_{j\ell}^i \omega^\ell,$$

$A_{j\ell}^i \equiv A_{\ell j}^i$.

Then we still have the formula

$$d\omega_i = \omega_j^i \wedge \omega^j,$$

and any (ω_j^i) which fits the formula must be given with suitable $A_{j\ell}^i$.

We claim that there is unique ω_j^i, as above with

$$\omega_j^i + \omega_i^j = 0.$$

Namely, pick one μ_j^i and write

$$\mu_j^i = \Sigma_k dw_k^i \wedge w_k^j + v_{j\ell}^i \omega^\ell.$$

Then $A_{j\ell}^i$ make ω_j^i satisfy the additional conditions, if and only if

$$v_{j\ell}^i + v_{i\ell}^j = A_{j\ell}^i + A_{i\ell}^j.$$

Now we play with the permutations of indexes. Since $A_{j\ell}^i = A_{\ell j}^i$,

$$v_{\ell j}^i + v_{ij}^\ell = A_{j\ell}^i + A_{ij}^\ell.$$

Hence

$$2A^i_{j\ell} = v^i_{j\ell} + v^i_{\ell j} + v^i_{i\ell} + v^\ell_{ij} - (A^j_{\ell i} + A^\ell_{ji})$$

$$= v^i_{j\ell} + v^i_{\ell j} + (v^j_{i\ell} - v^j_{\ell i}) + (v^\ell_{ij} - v^\ell_{ji}).$$

Thus, ω^i_j as above is unique if it exists. We then check, in fact, that for $A^i_{j\ell}$ as above, ω^i_j satisfies our requirement. It then follows

$$d\omega^i_j = \omega^i_k \wedge \omega^k_j + R^i_{j\ell k}\, \omega^\ell \wedge \omega^k.$$

$R^i_{j\ell k}\, \omega^\ell \wedge \omega^k$ is the curvature form of the Riemann geometry.

Note that in the above discussion the positivity of the metric was never used. What we need is to fix a non-degenerate symmetric quadratic form on R^n. (In the above we used the stand euclidian metric which gave rise to the skew-symmetric conditions on ω^i_j).

In what follows we give the construction of S. S. Chern and J. K. Moser [ ] for non-degenerate CR structures which follows very closely the above construction in Riemann geometry.

Before embarking on the construction we reformulate the definition of CR structure in terms of its defining equations. T'' will be defined (locally) by

$$d\nu = 0 \qquad (\nu = 1, \ldots, k)$$

where $d\nu \in C^\infty(\mathbb{C}T^*M)$. That is

$$X \in T'' \quad \text{if and only if} \quad \alpha_\nu(X) = 0 \qquad (\nu = 1, \ldots, n).$$

Note that for a Pfaffian form α

(3) $$d\alpha(X, Y) = X\alpha(Y) - Y\alpha(X) - \alpha([X, Y]).$$

Then we see that the condition: T'' is integrable is equivalent with the condition:

$$d\alpha_\nu(X, Y) = 0,$$

provided $\alpha_1(X) = \alpha_1(Y) = \ldots = \alpha_n(X) = \alpha_\nu(Y) = 0$, i.e., $d\alpha_\nu \equiv 0 \mod(\alpha_1, \ldots, \alpha_n)$.

The condition $T' \cap T'' = \{0\}$: note that $T' + T''$ is preserved under the operation of taking "bar". *Hence, it is the complexification of a subbundle* $T^0 \subset TM$ over R of codimension 1. Therefore

$$T^0 : \theta = 0 \qquad \theta = \overline{\theta}$$

for a suitable 1-form θ. Then we can find $\eta^1 \ldots \eta^{n-1}$ so that

$$T'' : \theta = \eta^1 = \ldots = \eta^{n-1} = 0$$

(because of the dimension $\theta = \eta^1 = \ldots = \eta^{n-1}$ must be linearly independent).

Then $T' : \theta = \overline{\eta}^1 = \ldots = \overline{\eta}^{n-1} = 0$. Therefore

$$T' \cap T'' = \{0\}$$

if and only if θ, $\eta^1 = \dots = \eta^{n-1}, \overline{\eta}^1, \dots, \overline{\eta}^{n-1}$ are linearly independent. Thus we find the following:

PROPOSITION. $T'' \subset \mathbb{C}TM$ *(*$\dim M = 2n-1$*) is a* CR *structure if and only if it is locally defined by*

$$T'' : \theta = \eta^1 = \dots = \eta^{n-1} = 0, \quad \theta = \overline{\theta}.$$

for suitable Pfaffian forms $\theta, \eta^1, \dots \eta^{n-1}$*, satisfying the following condition:*

1) $\theta, \eta^1, \dots, \eta^{n-1}, \overline{\eta}^{n-1}, \eta^{-1}, \dots, \overline{\eta}^{n-1}$ *are linearly independent.*

2) $d\theta \equiv 0$, $d\eta^i \equiv 0 \quad \mathrm{mod}(\theta, \eta^1, \dots, \eta^{n-1})$.

Note that $\eta^1, \dots, \eta^{n-1}$ restricted to T' form a base of $(T')^*$. To see Levi-form of T' in the above formulation, pick a real vector field s supplementary to T^0. Write

$$L(X, Y) \equiv L_s(X, Y)s \qquad \mathrm{mod}\, T'+T''$$

i.e.,

$$i[X, \overline{Y}] \equiv L_s(X, Y)s \qquad \mathrm{mod}\, T'+T''$$

(Since there is no intrinsic trivialization of the bundle $\mathbb{C}TM/(T' \oplus T'')$, L is represented by the set L_s. They differ only by multiplicative functions).

Note that, since $\theta = \overline{\theta}$, the above condition 2) implies that

(4)
$$d\theta \equiv i c_{j\overline{k}}\, \eta^j \wedge \overline{\eta}^k \wedge \overline{\eta}^k \quad \mathrm{mod}\, \theta$$

Then, for $X, Y \in C^\infty(T')$,

$$d\theta(X, \overline{Y}) = X\theta(\overline{Y}) - \overline{Y}\theta(X) - \theta([X, \overline{Y}])$$

$$= -\theta([X, \overline{Y}]) = iL_s(X, Y)\theta(S)$$

Therefore, we may write

(5)
$$L_s = \frac{1}{\theta(s)} c_{j\overline{k}}\, \eta^j \otimes \overline{\eta}^k$$

As remarked earlier, we fix a non-degenerate hermitian quadratic form

(4)
$$h_{j\overline{k}} z^j \overline{z}^k = \langle z, z \rangle,$$

$\overline{h_{j\overline{k}}} = h_{k\overline{j}}$ constants, $z = (z^1, \dots, z^{n-1}) \in \mathbb{C}^{n-1}$. We consider a CR structure T'' such that its Levi-form is similar to the above hermitian form, i.e., at each point in M there is an isomorphism $\iota: \mathbb{C}^{n-1} \longrightarrow T''$ such that $\langle z, z \rangle$ and $L_s(\iota(z), \iota(z))$ are proportional.

In the case of Riemann geometry we constructed the connection forms in two steps: ω^i first, then ω^i_j. In our case we need several steps. Set

(5) $$\tilde{E} = \{\theta \in T^*M: \theta(T^0) = 0, \ \theta \neq 0\}$$

We have the projection $\tilde{E} \longrightarrow M$ induced by the projection $T^*M \longrightarrow M$. Since T^0 is of co-dimension 1 in T, the set $\tilde{E}$ of non-zero annihilators of T^0 is a principal bundle where structure group is the multiplicative group $\mathbf{R}-\{0\}$. Thus, the fibres of $\tilde{E}$ have two components. In view of (4) and (5), we see the following: if for a local section $\theta(x)$ in a component of $\tilde{E}$, Levi-form is represented by $(c_{j\bar{k}})$, Levi-form is represented by $(-c_{j\bar{k}})$ in the other component. Hence, in one of the components, Levi-form is positive (because the positivity of $(c_{j\bar{k}})$ is a notion independent of choices of η_j). We call the elements in this component positive. It follows that $\tilde{E}$ has globally two components. Note also that for θ in the positive component, Levi-form (with respect to θ) has a representation $(h_{j\bar{k}})$.

§2. Let M be a CR structure such that its Levi-form is represented by $(h_{j\bar{k}})$. Then, as was shown in the first lecture, M may be covered by open sets U_0 together with linearly independent $\theta_0, \omega_0^1, \ldots, \omega_0^{n-1}$ (Pfaffian forms on U_0) satisfying the following conditions:

(0.1) $$\theta_0 = \omega_0^1 = \ldots = \omega_0^{n-1} = 0$$

defines T'' of the CR structure

(0.2) $$d\theta_0 = i h_{j\bar{k}} \omega_0^j \wedge \omega_0^{\bar{k}} + \theta_0 \wedge \pi_0$$
$$d\omega_0^k = \omega_0^k \wedge (\tau_0)_k^j + \theta_0 \wedge \varphi_0^j.$$

for a choice of Pfaffian forms $(\tau_0)_k^j$, φ_0^j.

We fix U_0, θ_0, ω_0^j, π_0, $(\tau_0)_k^j$, φ_0^j as above. They are used like a chart. We also note that $(\theta_0)_x$ is positive.

We set

$$E_1 = \{\theta \in T^*M;\ \theta = 0 \text{ defines } {}^\circ T_x \text{ (where } x \text{ is the source of } \theta) \text{ and } \theta \text{ positive}\}$$

Obviously E_1 is a subbundle of T^*M. We denote by $\rho_1 \colon E_1 \longrightarrow M$ the projection. Let Θ be the tautology form of E_1 over M. Namely, it is a Pfaffian form given by

$$\Theta_\theta = (\rho_1^* \theta)_\theta$$

where θ used as a suffix denotes the point θ in E_1, and θ in $\rho_1^* \theta$ denotes the element $\theta \in T^*M$. Clearly Θ is a Pfaffian form globally defined on E_1.

Over U_0, E_1 has a chart

(1)' $$U_0 \times \mathbb{R}^+ \ni (x, p) \longrightarrow p(\theta_0)_x \in E_1.$$

In terms of the above chart

(1) $$\Theta = p\theta_0$$

We set

$E_2 = \{(\theta, \omega);\ \theta \in E_1$ and $\omega = (\omega^1, ..., \omega^{n-1})$, where $\omega^j \in T^*_x M$ with $x = \rho_1\theta$, satisfying:
1) $\theta = \omega = 0$ defines T''_x
2) $(d\Theta)_\theta \equiv i h_{j\bar{k}}\, \omega^j \wedge \omega^{\bar{k}}$ mod $\Theta\}$

In the above we wrote ω^j instead of $(\rho_1^* \omega^j)_\theta$ for simplicity. We regard ω as $\mathbb{C}^{n-1}$-valued differential form at x. We see easily E_2 is a bundle over E_1 with the projection

$$\rho_2(\theta, \omega) = \theta.$$

In fact, we exhibit its chart later. Again we denote by Ω the tautology form of E_2 over E_1. Namely, $\Omega = (\Omega^1, ..., \Omega^{n-1})$ is a $\mathbb{C}^{n-1}$-valued Pfaffian form on E_2 given by

$$(\Omega^j)_{(\theta,\omega)} = (\rho_2^* \omega^j)_{(\theta,\omega)}$$

To write down a chart of E_2, we set

$$\langle \alpha, \beta \rangle = h_{j\bar{k}}\, \alpha^j \wedge \overline{\beta^k}$$

for $\mathbb{C}^{n-1}$-valued differential forms α, β. Then the condition 2) in the definition of E_2 is rewritten as

(2) $$(d\Theta)_\theta = i\langle \omega, \omega \rangle \qquad \text{mod } \Theta$$

For any $(\theta, \omega) \in E_2|U_0$ we see by the definition that we can write

(2)′ $$\theta = p\theta_0, \qquad p > 0$$

as before, and $$\omega = \omega_0 t + v\theta,$$

$t = (t^i_j) \in GL(n-1, \mathbb{C})$, $v = (v^1, ..., v^{n-1}) \in \mathbb{C}^{n-1}$, where ω_0 is regarded as *row* vector-valued, and $t = (t^i_j)$ as $(n-1)\times(n-1)$-matrix (with j indicating j-th row). Omitting ρ_1^* for simplicity we see

$$\begin{aligned} (d\Theta)_\theta &\equiv p\, d\theta_0 && \text{mod } \Theta \\ &\equiv p i\langle \omega_0, \omega_0 \rangle \\ &\equiv p i\langle \omega t^{-1}, \omega t^{-1} \rangle \equiv i\langle \omega(p t^{-1} t^{-1*}), \omega \rangle && \text{mod } \Theta \end{aligned}$$

Thus, (θ, ω) given by (2)′ is in E_2 if and only if $t^* t = pI$, where t^* denotes the conjugate of t with respect to the pairing $\langle z, z' \rangle$. Hence $E_2|U_0$ is a subbundle of $U_0 \times \mathbb{R}^+ \times GL(n-1, \mathbb{C}) \times \mathbb{C}^{n-1}$ given by the equation

(3) $$t^*t = pI$$

In this sense we may regard (x, p, t, v) as a chart of E_2, a chart with restraint. Note it is clear that E_2 is a bundle over E_1. In terms of the chart, we have

(4) $$\Omega = \omega_0 t + v\Theta.$$

Therefore $$d\Theta = pd\theta_0 + dp \wedge \theta_0 = ip\langle\omega_0, \omega_0\rangle + p\theta_0 \wedge \pi_0 + \Theta(-d\log p)$$

$$= ip\langle(\Omega - v\Theta)t^{-1}, (\Omega - v\Theta)t^{-1}\rangle + \Theta \wedge (-d\log p + \pi_0)$$

Hence, over U_0 we can write

(5) $$d\Theta = i\langle\Omega, \Omega\rangle + \Theta \wedge \pi$$
$$\pi = -d\log p + \pi_0 - i\langle v, \Omega\rangle + i\langle\Omega, v\rangle + s\Theta$$

where s can be any real number. By using a base including $\Omega^1, \ldots, \Omega^{n-1}, \Theta$ we see easily that the above π represent the all possible real-valued π which give the above expression of $d\Theta$. To make the above π unique (so that it is globally defined) we again use the idea of tautology form. Namely, we introduce

$$E = \{(\theta, \omega, \pi); (\theta, \omega) \in E_2 \text{ and } \pi \in T^*_{(\theta,\omega)}E_2 \text{ such that } (d\Theta)_{(\theta,\omega)} = i\langle\Omega, \Omega\rangle_{(\theta,\omega)} + (\Theta)_{(\theta,\omega)} \wedge \pi\}$$

Define the projection $$\rho_3\colon E \to (\theta, \omega, \pi) \longrightarrow (\theta, \omega) \in E_2$$

We introduce the tautology form Π of E over E_2. E has the fiber chart s (over E_2) introduced above, and we have the expression

(6) $$\Pi = -d\log p + \pi_0 - i\langle v, \Omega\rangle + i\langle\Omega, v\rangle + s\Theta$$

E has the chart (x, p, t, v, s). So far we have Θ, Ω, Π, Pfaffian forms on E_3 invariantly defined, with

(7) $$d\Theta = i\langle\Omega, \Omega\rangle + \Theta \wedge \Pi.$$

$\Theta, \Omega, \overline{\Omega}, \Pi$ generate dx, dp.

We look next at $d\Omega$. By (4) and (0)

$$d\Omega = -\omega_0 \wedge dt + (d\omega_0)\cdot t + dv \wedge \Theta + v(i\langle\Omega, \Omega\rangle + \Theta \wedge \Pi)$$
$$= (-\Omega + v\Theta)t^{-1} \wedge dt + (\Omega - v\Theta)t^{-1} \wedge \tau_0 t + \theta_0 \wedge \varphi_0 t + iv\langle\Omega, \Omega\rangle + \Theta \wedge (-dv + v\Pi)$$
$$= \Omega \wedge (-t^{-1}dt + t^{-1}\tau_0 t) + iv\langle\Omega, \Omega\rangle + \Theta \wedge (-dv + v\Pi + vt^{-1} \wedge dt - vt^{-1}\tau_0 t + \frac{1}{p}\varphi_0 t)$$

Hence, on U_0

$$d\Omega = \Omega \wedge \tau + \Theta \wedge \varphi$$

(8) $$\tau^j_k = -(t^{-1}dt)^j_k + (t^{-1}\tau_0 t)^j_k + iv^j h_{k\bar{\ell}} \overline{\Omega}^\ell + (\tau_1)^i_j$$

$$\varphi = -dv + v\Pi + vt^{-1} \wedge dt - vt^{-1}\tau_0 t + \frac{1}{p}\varphi_0 t + \varphi_1$$

where $(\tau_1)^j_k$ and $(\varphi_1)^j$ are any Pfaffian forms satisfying

(9) $$\Omega \wedge \tau_1 + \Theta \wedge \varphi_1 = 0.$$

We have to impose conditions on τ and φ so that they are uniquely determined. τ corresponds to (ω^i_j) in the case of Riemann geometry. (ω^i_j) was uniquely determined by setting the conditions $\omega^i_j + \omega^j_i = 0$. Note that the bundle of the orthonormal coframes has the fiber chart $g \in O(n)$ and the restriction of ω^i_j to the fibers are Mauer-Cartan forms $(g^{-1}dg)^i_j$. Thus the condition we imposed was the condition satisfied by Mauer-Cartan form. Similarly we have a fiber chart (t, p) with the condition (3) and the restrictions to the fibers of τ and Π are

(10) $$-t^{-1}dt \quad \text{and} \quad -d\log p$$

(because we see by (9) that τ_1 and φ_1 are zero modulo $\Omega^1, \ldots, \Omega^{n-1}$, and Θ).

Thus we may guess that the condition we should impose on τ is the symmetry condition satisfied by the elements in (10) derived from (3).

By taking the exterior derivative of (3) we see

$$t^*dt + (t^*dt)^* = dp I$$

Since $t^* = pt^{-1}$ it then follows that

$$t^{-1}dt + (t^{-1}dt)^* = (d\log P) I$$

In view of (9) it seems reasonable to try the condition

(11) $$\tau + \tau^* - \Pi I = 0,$$

where τ^* is defined as the $(n-1)\times(n-1)$-matrix-valued differential form satisfying the condition

$$\langle z, z'\tau\rangle = \langle z\tau^*, z'\rangle$$

for any complex row $(n-1)$-vectors z, z'.

We claim that we can actually find such τ for a suitable φ. Namely, pick any τ and φ on U_0 satisfying (8) and denote them by τ, φ. Then, by taking d of (7)

$$0 = i\langle d\Omega, \Omega\rangle - i\langle \Omega, d\Omega\rangle + d\Theta \wedge \Pi - \Theta \wedge d\Pi$$

$$= i\langle\Omega \wedge \tilde{\tau}, \Omega\rangle - i\langle\Omega, \Omega \wedge \tilde{\tau}\rangle + i\langle\Theta \wedge \tilde{\varphi}, \Omega\rangle - i\langle\Omega, \Theta \wedge \tilde{\varphi}\rangle + i\langle\Omega, \Omega\rangle \wedge \Pi - \Theta \wedge d\Pi$$

$$= i\langle\Omega \wedge \tilde{\tau} + \Omega \wedge (\tilde{\tau})^* - \Omega \wedge \Pi, \Omega\rangle + \Theta \wedge (-d\Pi + i\langle\tilde{\varphi}, \Omega\rangle + i\langle\Omega, \tilde{\varphi}\rangle).$$

Hence

$$i\langle\Omega \wedge \alpha, \Omega\rangle + \Theta \wedge \beta = 0,$$

where

(12)
$$\alpha = \tau + \tau^* - \Pi I$$
$$\beta = -d\Pi + i\langle\varphi, \Omega\rangle + i\langle\Omega, \varphi\rangle$$

Since $\Omega^1, \ldots, \Omega^{n-1}, \overline{\Omega^1}, \ldots, \overline{\Omega^{n-1}}, \Theta$ are linearly independent, the relation implies that α and β are zero modulo $\Omega^1, \ldots, \Omega^{n-1}, \overline{\Omega^1}, \ldots, \overline{\Omega^{n-1}}, \Theta$. Besides, $\alpha = \alpha^*$ and $\beta = \overline{\beta}$. Therefore we can write

$$\alpha = \Omega^j A_j + \overline{\Omega^j} A_j^* + \Theta A_0$$

where A_0, A_j are matrix-valued functions on U_0 with $A_0^* = A_0$. To see the implications of the relations (12) on A_0, A_j, it is convenient to call a differential form type (p, q) when it is a sum of terms containing exactly p factors of Ω^i's and q factors of $\overline{\Omega^j}$'s. Then, looking at terms of type (2,1), we find that the relation implies that

$$\Omega \wedge (\Omega^j A_j) = 0,$$

When we set

$$\tau = \tilde{\tau} - \Omega^j A_j - \frac{1}{2}\Theta A_0$$
$$\varphi = \tilde{\varphi} - \frac{1}{2}\Omega A_0,$$

we find that

(13)
$$d\Omega = \Omega \wedge \tau + \Theta \wedge \varphi$$
$$\tau + \tau^* - \Pi I = 0$$

τ and φ as above are not unique. However, the following hold:

If τ_1 and φ_1 are as above, then

$$\tau - \tau_1 = \Theta S_0$$
$$\varphi - \varphi_1 \equiv \Omega S_0 \qquad \text{mod } \Theta$$

for a matrix-valued function S_0 with $S_0 + S_0^* = 0$ on $\rho^{-1}(U_0)$, and viceversa. In fact, set

$$S = \tau - \tau_1, \qquad B = \varphi - \varphi_1.$$

Then we have

$$\Omega \wedge S + \Theta \wedge B = 0, \qquad S + S^* = 0.$$

Write

$$S = \Omega^j S_j - \overline{\Omega}^j S_j^* + \Theta S_0$$

where S_0 and S_j are matrix-valued functions, and $S_0^* + S_0 = 0$. Then, looking at terms of type (1, 1) in the equality $\Omega S + \Theta B = 0$, we find easily that $S_j = 0$. Then it follows that $\Theta \wedge (B - \Omega S_0) = 0$, i.e., $B \equiv \Omega S_0 \mod \Theta$.

We pick τ, φ on $\rho^{-1}(U_0)$ satisfying (13). Then (12) is still valid when $\tilde{\tau}$ and $\tilde{\varphi}$ happen to be our τ and φ. In this case $\alpha = 0$ and hence $\Theta\beta = 0$. This means we can write

$$d\Pi = i\langle\varphi, \Omega\rangle + i\langle\Omega, \varphi\rangle + \Theta \wedge \psi$$

for a Pfaffian form ψ on $\rho^{-1}(U_0)$. When we change τ by adding ΘS_0 (with $S^* + S = 0$), we have to add ΩS_0 to φ to keep our structure equation. However, this does not change the term $\langle\varphi, \Omega\rangle + \langle\Omega, \varphi\rangle$. Therefore, the above defining formula of ψ implies that ψ is uniquely determined modulo Θ. By (6) we find that

$$\psi \equiv -ds \qquad \mod(\Theta, \Omega, \tau, \varphi).$$

Our result so far is summarized as follows: we have uniquely and globally defined Θ, Ω, Π, on E with locally defined τ, φ, ψ forming locally a base (under the restraint condition $\tau + \tau^* = \Pi I$) of Pfaffian forms on E satisfying the following conditions:

$$\begin{aligned} d\Theta &= i\langle\Omega, \Omega\rangle + \Theta \wedge \Pi, \quad \Theta = \overline{\Theta}, \quad \Pi = \overline{\Pi} \\ d\Omega &= \Omega \wedge \tau + \Theta \wedge \varphi, \qquad \tau + \tau^* - \Pi I = 0 \\ d\Pi &= i\langle\varphi, \Omega\rangle + i\langle\Omega, \varphi\rangle + \Theta \wedge \psi, \qquad \psi = \overline{\psi} \end{aligned} \tag{14}$$

If (τ, φ, ψ) satisfies (14), any other choice is of the form

$$\begin{aligned} \tilde{\tau} &= \tau + \Theta D, \qquad D + D^* = 0 \\ \tilde{\varphi} &= \varphi + \Omega D + \Theta V \\ \tilde{\psi} &= \psi + i\langle\Omega, V\rangle - i\langle V, \Omega\rangle - \Theta F, \qquad F = \overline{F} \end{aligned} \tag{15}$$

where D is matrix-valued, V is C^{n-1}-valued, F is a real-valued function.

Now we exhausted the analogy with Riemann geometry. Because in the case of Riemann geometry we do not have the direction Θ. And the above summary shows that all the non-uniqueness stems from that direction. To make τ, φ, ψ unique so that we have a globally defined form we need a new idea. The idea is to calculate the curvature forms locally and to put conditions on them so that we obtain locally unique τ, φ, ψ. This means then that τ, φ, ψ will be defined globally.

§**3**. Starting from a CR structure (with non-degenerate Levi form) we constructed the bundle E over M as well as Θ, Ω, Π, in a concrete way. However, in the following discussion all we need is the relation (2.14). Hence we start from the following situation: Let E be a manifold. Assume that we are given Pfaffian forms Θ, Ω^j $(j = 1, \ldots, n-1)$, Π on E such that with locally defined $\tau = (\tau^j_k)$, $\varphi = (\varphi^i)$, and ψ we have the formula (2.14). Assume further that $\Theta, \Omega^i, \overline{\Omega^j}, \Pi, \tau^i_k, \varphi^j, \psi$ form a base of complex-valued Pfaffian forms of E under the sole relation $\tau + \tau^* = \Pi I$.

Actually any such (E, Θ, Ω, Π) is locally isomorphic to the one constructed in §2. However, not to interrupt our construction, we postpone this discussion until later. Similarly, the structure of the fibers of E will be treated later.

Taking d of the formula for $d\Omega$ in (2.14), we find that

$$0 = d\Omega \wedge \tau - \Omega \wedge d\tau + d\Theta \wedge \varphi - \Theta \wedge d\varphi$$

$$= (\Omega \wedge \tau + \Theta \wedge \varphi) \wedge \tau - \Omega \wedge d\tau + (i\langle \Omega, \Omega \rangle + \Theta \wedge \Pi) \wedge \varphi - \Theta \wedge d\varphi$$

Hence

$$\Omega \wedge \alpha + \Theta \wedge \beta = 0$$

where

(1)
$$\alpha^i_k = -d\tau^j_k + \tau^\ell_k \wedge \tau^j_\ell + i h_{k\overline{\ell}} \overline{\Omega^\ell} \wedge \varphi^j$$
$$\beta = -d\varphi + \varphi \wedge \tau + \Pi \wedge \varphi$$

Thus we have $\Omega \wedge \alpha \equiv 0 \mod \Theta$. However, since α is a 2-form, all we get out of the information is that $\alpha = \alpha_j \wedge \Omega^j \mod \Theta$, with $(\alpha_j)^s_k = (\alpha_k)^s_j$. However, if moreover $\Omega \wedge \alpha^* \equiv 0 \mod \Theta$, then we can also write $\alpha = \alpha'_i \wedge \overline{\Omega^j} \mod \Theta$. Therefore $\alpha \equiv A_{j\overline{k}} \Omega^j \wedge \overline{\Omega^k} \mod \Theta$. With a hope of obtaining this type of information, we calculate α^*. (Actually things do not turn out as we hope and we have to make an adjustment).

Since

(2)
$$z\alpha = -zd\tau + z\tau \wedge \tau + i\langle z, \Omega \rangle \wedge \varphi,$$

$$\langle z\alpha, w \rangle = -\langle zd\tau, w \rangle + \langle z\tau \wedge \tau, w \rangle + i\langle z, \Omega \rangle \wedge \langle \varphi, w \rangle$$
$$= -\langle \tau, w(d\tau)^* \rangle - \langle z, w\tau^* \wedge \tau^* \rangle - \langle z, i\Omega \wedge \langle w, \varphi \rangle \rangle$$

Therefore

$$
\begin{aligned}
w\alpha^* &= -w d\tau^* - w\tau^* \wedge \tau^* - i\Omega \wedge \langle w, \varphi\rangle \\
&= w(d\tau - d\Pi) - w(\tau - \Pi I) \wedge (\tau - \Pi I) + i\langle w, \varphi\rangle \wedge \Omega \\
&= w(d\tau - \tau \wedge \tau) - w d\Pi + w\tau \wedge \Pi + \Pi \wedge w\tau + i\langle w, \varphi\rangle \wedge \Omega \\
&= -w\alpha + i\langle w, \Omega\rangle \wedge \varphi + i\langle w, \varphi\rangle \wedge \Omega - w d\Pi \\
&= -w\alpha + i\langle w, \Omega\rangle \wedge \varphi + i\langle w, \varphi\rangle \wedge \Omega - iw\langle \Omega, \varphi\rangle - iw\langle \varphi, \Omega\rangle - \Theta \wedge \psi
\end{aligned}
$$

Thus we find to our dismay that $\Omega \wedge (\alpha+\alpha^*) \not\equiv 0 \mod \Theta$. To study the situation more carefully, we write out the terms in $\alpha+\alpha^*$, which does not cause us trouble. Namely,

(3) $$w(\alpha+\alpha^*) = -w\Theta \wedge \psi + i\langle w, \Omega\rangle \wedge \varphi - iw\langle \varphi, \Omega\rangle + w\gamma$$

where

(3)′ $$w\gamma = i\langle w, \varphi\rangle \wedge \Omega - iw\langle \Omega, \varphi\rangle$$

so that

$$\Omega \wedge \gamma = 0.$$

Now, calculating as before we find easily that

$$z\gamma^* = -i\varphi \wedge \langle z, \Omega\rangle + iz\overline{\langle \Omega, \varphi\rangle}$$

However, since

$$\langle \Omega, \varphi\rangle = h_{j\bar{k}}\, \Omega^j \wedge \overline{\varphi^k}$$

we find that $\overline{\langle \Omega, \varphi\rangle} = -\langle \varphi, \Omega\rangle$. Therefore

$$w\gamma^* = i\langle w, \Omega\rangle \wedge \varphi - iw\langle \varphi, \Omega\rangle$$

so that

$$w(\alpha+\alpha^*) = -w\Theta \wedge \psi + w(\gamma+\gamma^*).$$

Now we see the light at last. Namely, if we set

(4) $$w\delta = w\alpha + \frac{1}{2} w\Theta \wedge \psi - w\gamma$$

then we see

$$\Omega \wedge \delta + \Theta \wedge \epsilon = 0,$$

(5) $$\delta + \delta^* = 0,$$

where

(6) $$\epsilon = -d\varphi + \varphi \wedge \tau + \Pi \wedge \varphi + \frac{1}{2}\Omega \wedge \psi$$

As was seen in the discussion following the formula (1), (5) implies that

(7) $$\delta \equiv -R^{\tau}_{j\bar{k}}\Omega^j \wedge \overline{\Omega^k} \qquad \mathrm{mod}\ \Theta$$

where $R^{\tau}_{j\bar{k}}$ is a matrix-valued function on U_0. Since

$$\langle z\delta, w\rangle = -\langle zR^{\tau}_{j\bar{k}}\, w\rangle \Omega^j \wedge \overline{\Omega^k} = \langle z, (-1) w (R^{\tau}_{jk})^* \overline{\Omega^j} \wedge \Omega^k\rangle,$$

we see that $\delta^* = (R^{\tau}_{jk})^* \Omega^k \wedge \overline{\Omega^j}$. Hence we find by (5)

(8) $$(R^{\tau}_{j\bar{k}})^* = R^{\tau}_{k\bar{j}}, \qquad (R^{\tau}_{j\bar{k}})^s_{\ell} + (R^{\tau}_{\ell\bar{k}})^s_j = 0.$$

When we write $\delta = -R^{\tau}_{j\bar{k}}\Omega^j \wedge \overline{\Omega^k} - \Theta \wedge \lambda$ we see by (5), (7) and (8) that $\Theta \wedge (\epsilon + \Omega \wedge \lambda) = 0$, i.e. $\epsilon \equiv \lambda \wedge \Omega \quad \mathrm{mod}\ \Theta$. We pause here to summarize what we found so far:

(9) $$z d\tau = \dot{z}\tau \wedge \tau + i\langle z, \Omega\rangle \wedge \varphi - i\langle z, \varphi\rangle \wedge \Omega + iz\langle \Omega, \varphi\rangle + \frac{1}{2} z\Theta \wedge \psi + zR^{\tau},$$

where

$$R^{\tau} = R^{\tau}_{j\bar{k}}\Omega^j \wedge \overline{\Omega^k} + \Theta \wedge \lambda$$

(Cf. (2), (4), and (3)′)

(10) $$d\varphi \equiv \varphi \wedge \tau + \Pi \wedge \varphi + \frac{1}{2}\Omega \wedge \psi + \Omega \wedge \lambda \qquad \mathrm{mod}\ \Theta$$

Following our program we next find out how R^{τ} changes when we change τ as in (2.15). Set $\tilde{\tau} = \tau + \Theta D$ where $D + D^* = 0$. Then we have to change φ to $\tilde{\varphi} \equiv \varphi + \Omega D \quad \mathrm{mod}\ \Theta$. Calculating $\mathrm{mod}\ \Theta$ we find by (9)

$$\begin{aligned} \tilde{z} d\tau &\equiv z d\tau + D d\Theta \\ &\equiv z\tilde{\tau} \wedge \tilde{\tau} + i\langle z, \Omega\rangle \wedge (\tilde{\varphi} - \Omega D) - i\langle z, \tilde{\varphi} - \Omega D\rangle \wedge \Omega + iz\langle \Omega, \tilde{\varphi} - \Omega D\rangle + zR^{\tau} + Di\langle \Omega, \Omega\rangle \\ &\equiv z\tilde{\tau} \wedge \tilde{\tau} + i\langle z, \Omega\rangle \wedge \tilde{\varphi} - i\langle z, \tilde{\varphi}\rangle \wedge \Omega + iz\langle \Omega, \tilde{\varphi}\rangle + zR^{\tau} - i\langle z, \Omega\rangle \wedge \Omega D + i\langle z, \Omega D\rangle \wedge \Omega \\ &\qquad - iz\langle \Omega, \Omega D\rangle + Di\langle \Omega, \Omega\rangle. \end{aligned}$$

Therefore

$$zR^{\tilde{\tau}} \equiv zR^{\tau} - i\langle z, \Omega\rangle \wedge \Omega D - i\langle zD, \Omega\rangle \wedge \Omega + iz\langle \Omega D, \Omega\rangle + izD\langle \Omega, \Omega\rangle \qquad \mathrm{mod}\ \Theta.$$

These are matrix-valued 2-forms. For a matrix $A = (A^j_k)$ we set

$$\operatorname{tr} A = A^j_j$$

When we introduce a matrix $(h^{j\bar{k}})$ by

$$h^{j\bar{\ell}} h_{k\bar{\ell}} = \delta^j_k$$

we see that

(11)
$$\operatorname{tr} A = h^{j\bar{k}} \, \partial^2 \langle zA, z\rangle / \partial z^j \partial \bar{z}^k$$

Since

$$\langle z\widetilde{R^\tau}, z\rangle \equiv \langle zR^\tau, z\rangle - i\langle z, \Omega\rangle \wedge \langle \Omega D, z\rangle - i\langle zD, \Omega\rangle \wedge \langle \Omega, z\rangle + i\langle z, z\rangle \langle \Omega D, \Omega\rangle + i\langle zD, z\rangle \langle \Omega, \Omega\rangle \mod \Theta,$$

$$\operatorname{tr} \widetilde{R^\tau} = \operatorname{tr} R^\tau - i\overline{\Omega^k} \wedge \langle \Omega D, e_k\rangle - i\langle e_j D, \Omega\rangle \wedge \Omega^j + i(n-1)\langle \Omega D, \Omega\rangle + i(\operatorname{tr} D)\langle \Omega, \Omega\rangle \mod \Theta$$

where $\{e_1, \ldots, e_{n-1}\}$ is the standard base of $\mathbb{C}^{n-1}$, i.e.,

$$\operatorname{tr} \widetilde{R^\tau}_{j\bar{k}} = \operatorname{tr} R^\tau_{j\bar{k}} + i(n+1) h_{\ell\bar{k}} D^\ell_j + i(\operatorname{tr} D) h_{j\bar{k}}$$

We now claim that there is unique D such that $\operatorname{tr} \widetilde{R^\tau}_{j\bar{k}} = 0$. Namely, for any such D

$$i \operatorname{tr} R^\tau_{j\bar{k}} = (n+1) h_{\ell\bar{k}} D^\ell_j + (\operatorname{tr} D) h_{j\bar{k}}$$

i.e.,
$$i \operatorname{tr}(h^{j\bar{k}} R^\tau_{j\bar{k}}) = 2n \operatorname{tr} D.$$

Hence

$$h_{\ell\bar{k}} D^\ell_j = \frac{1}{n+1} \operatorname{tr} R^\tau_{j\bar{k}} - \frac{i}{2n(n+1)} h_{j\bar{k}} \operatorname{tr}(h^{s\bar{t}} R^\tau_{s\bar{t}}),$$

i.e.

(12)
$$D^\ell_j = \frac{i}{n+1} \operatorname{tr}(h^{\ell\bar{k}} R^\tau_{j\bar{k}}) - \frac{i}{2n(n+1)} \delta^\ell_j \operatorname{tr}(h^{s\bar{t}} R^\tau_{s\bar{t}}).$$

Therefore, such D is unique if it exists.

Since

$$\langle wD, w\rangle = \frac{i}{n+1} \operatorname{tr}(R^\tau_{j\bar{k}}) w^j \overline{w^k} - \frac{i}{2n(n+1)} \langle w, w\rangle h^{s\bar{t}} \operatorname{tr}(R^\tau_{s\bar{t}})$$

it follows immediately that $\langle wD, w\rangle$ is purely imaginary, i.e., $D = -D^*$. Now we can verify directly that D given by the above formula satisfies our requirement.

Now we can conclude that there is a unique globally-defined Pfaffian form τ and locally-defined

φ, ψ, and λ such that

$$z d\tau = z\tau \wedge \tau + i\langle z, \Omega\rangle \wedge \varphi - i\langle z, \varphi\rangle \wedge \Omega + iz\langle \Omega, \varphi\rangle + \frac{1}{2} z\Theta \wedge \psi + zR^{\tau}.$$

(13)
$$\tau + \tau^* - \Pi I = 0$$

$$R^{\tau} = R^{\tau}_{j\bar{k}}\Omega^j \wedge \overline{\Omega^k} + \Theta \wedge \lambda$$

$$\operatorname{tr}(R^{\tau}_{j\bar{k}}) = 0, \qquad (R^{\tau}_{j\bar{k}})^* = R^{\tau}_{k\bar{j}}, \qquad (R^{\tau}_{j\bar{k}})^s_{\ell} + (R^{\tau}_{\ell\bar{k}})^s_j = 0$$

(14)
$$d\varphi \equiv \varphi \wedge \tau + \Pi \wedge \varphi + \frac{1}{2}\Omega \wedge \psi + \Omega \wedge \lambda \qquad \text{mod } \Theta.$$

Any other choice of φ, ψ is of the form

(15)
$$\tilde{\varphi} = \varphi + \Theta V$$

$$\tilde{\psi} = \psi + i\langle \Omega, V\rangle - i\langle V, \Omega\rangle + \Theta F$$

Note that undetermined φ, ψ enter in the formula for $d\tau$ and hence R^{τ} may depend on them. Since they are determined mod Θ, we see that $R^{\tau}_{j\bar{k}}$ is completely determined. However, λ will depend on the choices of φ.

We will use a similar argument to make the choice of φ and ψ unique.

§4. We should obtain more information on λ out of the relation: $\tau + \tau^* = \Pi I$ and $d(d\tau) = 0$. Namely, by taking $*$ of (3.13), we see

$$\begin{aligned}
z d\tau^* &= -z\tau^* \wedge \tau^* - i\Omega \wedge \langle z, \varphi\rangle + i\varphi \wedge \langle z, \Omega\rangle - i\overline{\langle \Omega, \varphi\rangle} z + \frac{1}{2} z\Theta \wedge \psi + z(R^{\tau})^* \\
&= -z(\tau - \Pi I) \wedge (\tau - \Pi I) - i\Omega \wedge \langle z, \varphi\rangle + i\varphi \wedge \langle z, \Omega\rangle + iz\langle \varphi, \Omega\rangle + \frac{1}{2} z\Theta \wedge \psi + z(R^{\tau})^* \\
&= -z\tau \wedge \tau + (z\tau \wedge \Pi + \Pi \wedge z\tau) + i\langle z, \varphi\rangle \wedge \Omega - i\langle z, \Omega\rangle \wedge \varphi + iz\langle \varphi, \Omega\rangle + \frac{1}{2}\Theta \wedge \psi + z(R^{\tau})^* \\
&= -z d\tau + iz\langle \Omega, \varphi\rangle + iz\langle \varphi, \Omega\rangle + z\Theta \wedge \psi + z(R^{\tau})^* + zR^{\tau} \\
&= -z d\tau + z d\Pi + z(R^{\tau} + (R^{\tau})^*).
\end{aligned}$$

Hence $R^{\tau} + (R^{\tau})^* = 0$. In view of the relation in $R^{\tau}_{j\bar{k}}$ in (3.13) it then follows that

(1)
$$\lambda + \lambda^* = 0.$$

Since $d(d\tau) = 0$

$$0 \equiv z d\tau \wedge \tau - z\tau \wedge d\tau + i\langle z, d\Omega\rangle \wedge \varphi - i\langle z, \Omega\rangle \wedge d\varphi - i\langle z, d\varphi\rangle \wedge \Omega + i\langle z, \varphi\rangle \wedge d\Omega + iz\langle d\Omega, \varphi\rangle - iz\langle \Omega, d\varphi\rangle$$

$$+\frac{1}{2} z d\Theta \quad \psi + z dR^{\tau} \qquad \text{mod } \Theta.$$

We replace $d\tau, d\Omega \; d\varphi, d\Theta$ in the above by their expression, and observe the type (1,1)-part mod Θ. We find that

$$\text{(2)} \qquad 0 \equiv z(R^{\tau})^{(1,1)} \wedge \tau - z\tau \wedge (R^{\tau})^{(1,1)} - i\langle z, \Omega\rangle \wedge \Omega \wedge (\frac{1}{2}\psi + \lambda) - i\langle z, \Omega \wedge (\frac{1}{2}\psi + \lambda)\rangle \wedge \Omega$$

$$- iz\langle \Omega, \Omega \wedge (\frac{1}{2}\psi + \lambda)\rangle + \frac{i}{2} z\langle \Omega, \Omega\rangle \wedge \psi + z(dR^{\tau})^{(1,1)} \qquad \text{mod } \Theta$$

$$\equiv z(R^{\tau})^{(1,1)} \wedge \tau - z\tau \wedge (R^{\tau})^{(1,1)} - i\langle z, \Omega\rangle \wedge \Omega \wedge \lambda - i\langle z, \Omega \wedge \lambda\rangle \wedge \Omega - iz\langle \Omega, \Omega \wedge \lambda\rangle + z(dR^{\tau})^{(1,1)}$$

On the other hand, by the formula for R^{τ} in (3.13)

$$(dR^{\tau})^{(1,1)} \equiv (dR^{\tau}_{j\bar{k}})^{(0,0)} \wedge \Omega^j \wedge \overline{\Omega^k} + R^{\tau}_{\ell\bar{k}}(\Omega \wedge \tau)^{\ell} \wedge \overline{\Omega^k} - R^{\tau}_{j\bar{\ell}} \Omega^j \wedge \overline{(\Omega \wedge \tau)^k} + i\langle \Omega, \Omega\rangle \wedge \lambda \qquad \text{mod } \Theta$$

$$\equiv ((dR^{\tau}_{j\bar{k}})^{(0,0)} - R^{\tau}_{\ell\bar{k}} \tau^{\ell}_j - R^{\tau}_{j\bar{\ell}} \overline{\tau^{\ell}_k} + i h_{j\bar{k}} \lambda) \Omega^j \wedge \overline{\Omega^k} \qquad \text{mod } \Theta.$$

Note that $R^{\tau}_{j\bar{k}}, \lambda$ are matrix-valued. We take the trace of their value. Then, by the trace condition of $R^{\tau}_{j\bar{k}}$

$$\text{tr}(dR^{\tau})^{(1,1)} \equiv i(\text{tr}\,\lambda)\langle \Omega, \Omega\rangle \qquad \text{mod } \Theta.$$

Hence, applying the trace to the value in (2) and noting $\text{tr}(AB) = \text{tr}(BA)$, we find (Cf. (3.11)) that ($e_1, \ldots, e_n$ denote the standard base of $\mathbb{C}^{n-1}$)

$$0 \equiv -i\overline{\Omega^k} \wedge \langle \Omega \wedge \lambda, e_k\rangle - i\langle \overline{(\Omega \wedge \lambda)^k} \wedge \langle \Omega, e_k\rangle - i(n-1)\langle \Omega, \Omega \wedge \lambda\rangle + i(\text{tr}\,\lambda)\langle \Omega, \Omega\rangle \qquad \text{mod } \Theta$$

$$= -i\langle \Omega \wedge \lambda, \Omega\rangle - in\langle \Omega, \Omega \wedge \lambda\rangle + i(\text{tr}\,\lambda)\langle \Omega, \Omega\rangle$$

$$= -i(n+1)\langle \Omega \wedge \lambda, \Omega\rangle + i(\text{tr}\,\lambda)\langle \Omega, \Omega\rangle$$

by (1). Therefore

$$(n+1) h_{\ell\bar{k}} \lambda^{\ell}_j + h_{j\bar{k}} \text{tr}\,\lambda \equiv 0 \qquad \text{mod } \Omega, \overline{\Omega}, \Theta$$

i.e.

$$\lambda \equiv 0 \qquad \text{mod } \Omega, \overline{\Omega}, \Theta.$$

Therefore, in view of (1)

$$\text{(3)} \qquad \lambda \equiv R^{\tau}_j \Omega^j + R^{\tau}_{\bar{k}} \Omega^{\bar{k}} \quad \text{mod } \Theta, \qquad R^{\tau}_{\bar{k}} = -(R^{\tau}_k)^*$$

where R^{τ}_j is a locally defined matrix-valued function.

We examine the change in λ due to a change of (φ, Ψ) to $(\tilde{\varphi}, \tilde{\Psi})$ as in (3.15). By (3.14)

$$d\tilde{\varphi} \equiv d\varphi + V d\Theta \qquad \text{mod } \Theta$$

$$\equiv \varphi \wedge \tau + \Pi \wedge \varphi + \frac{1}{2}\Omega \wedge \Psi + \Omega \wedge \lambda + i\langle \Omega, \Omega \rangle V \mod \Theta$$

$$\equiv \widetilde{\varphi} \wedge \tau + \Pi \wedge \widetilde{\varphi} + \frac{1}{2}\Omega \wedge (\widetilde{\Psi} - i\langle \Omega, V\rangle + i\langle V, \Omega\rangle) + \Omega \wedge \lambda + i\langle \Omega, \Omega \rangle V \mod \Theta.$$

Therefore

$$\Omega \wedge \widetilde{\lambda} = \Omega \wedge (\lambda - i\langle \Omega, V\rangle + i\langle V, \Omega\rangle) + i\langle \Omega, \Omega\rangle V.$$

Since $\Omega, \overline{\Omega}$ are independent, it then follows that

$$z(\widetilde{\lambda})^{(0,1)} = z(\lambda)^{(0,1)} + i\langle V, \Omega\rangle z + i\langle z, \Omega\rangle V.$$

Applying * and using (1) we see

$$z(\widetilde{\lambda})^{(1,0)} = z(\lambda)^{(0,1)} + i\langle \Omega, V\rangle z + i\langle z, V\rangle \Omega$$

i.e.,
$$(\widetilde{R}^{\tau}_j)^t_s = (R^{\tau}_j)^t_s + i\delta^t_s h_{j\bar{k}} \overline{V^k} + i\delta^t_j h_{s\bar{k}} \overline{V^k}$$

Now we see that we can impose conditions on $(\widetilde{R}^{\tau}_j)^t_s$ so that V is unique. We may require $(\widetilde{R}^{\tau}_j)^j_s = 0$, or $(\widetilde{R}^{\tau}_j)^t_s = 0$.

However, the one which makes the subsequent arguments work is the first one. Thus we now have a globally defined Φ such that

(4)
$$z d\tau = z\tau \wedge \tau + i\langle z, \Omega\rangle \wedge \Phi - i\langle z, \Phi\rangle \wedge \Omega + iz\langle \Omega, \Phi\rangle + \frac{1}{2} z\Theta \wedge \psi + zR^{\tau}$$

(5)
$$R^{\tau} = R^{\tau}_{j\bar{k}} \Omega^j \wedge \overline{\Omega^k} + \Theta \wedge \lambda$$

$$\lambda = R^{\tau}_j \Omega^j + R^{\tau}_{\bar{k}} \overline{\Omega^k}, \qquad R^{\tau}_{\bar{k}} = -(R^{\tau}_k)^*$$

(6)
$$(R^{\tau}_{j\bar{k}})^* = R^{\tau}_{k\bar{j}}, \qquad (R^{\tau}_{j\bar{k}})^t_{\ell} = (R^{\tau}_{\ell\bar{k}})^t_j$$

$$\operatorname{tr} R^{\tau}_{j\bar{k}} = 0, \qquad (R^{\tau}_j)^j_{\ell} = 0$$

(7)
$$d\Phi = \Phi \wedge \tau + \Pi \wedge \Phi + \frac{1}{2}\Omega \wedge \psi + R^{\Phi}$$

$$R^{\Phi} = \Omega \wedge \lambda + \Theta \wedge \mu$$

It remains to show that $\mu \equiv 0 \mod \Omega, \overline{\Omega}, \Theta$, and to make the choice of ψ unique, and to determine the expression of $d\psi$. We first find an expression of $d\psi$. Namely, we calculate $d(d\Pi)$ by (2.14) where φ is now replaced by Φ. Considering all terms containing Θ,

$$0 = i\Theta \wedge \langle \mu, \Omega\rangle - i\langle \Phi, \Theta \wedge \Phi\rangle + i\langle \Theta \wedge \Phi, \Phi\rangle - i\langle \Omega, \Theta \wedge \mu\rangle + \Theta \wedge \Pi \wedge \psi - \Theta \wedge d\psi,$$

i.e.,

$$d\psi = \Pi \wedge \psi + 2i\langle \Phi, \Phi\rangle + R^{\psi},$$

(8)

$$R^{\psi} = i\langle \mu, \Omega\rangle + i\langle \Omega, \mu\rangle + \Theta \wedge \gamma$$

for a locally-defined Pfaffian form γ.

Before proceeding further, we work out, since we need them, some consequences of the conditions (5), (6) on R^{τ}_{k}, $R^{\tau}_{\bar k}$, $R^{\tau}_{j\bar k}$. They are

(9) $$R^{\tau}_{j\bar k} h^{j\bar k} = 0$$

(10) $$(R^{\tau}_{\bar k})^{t}_{j}\, h^{j\bar k} = 0$$

(11) $$(\tau_j^{\ell}(R^{\tau}_{\bar k} + (R^{\tau}_{\bar\ell})^{t}_{j}\, \overline{\tau_k^{\ell}})\, h^{j\bar k} = 0$$

To prove these, note that, for a matrix A, $h^{\ell\bar k}(A^*)^{t}_{\ell} = \overline{A^{k}_{\ell}}\, h^{t\bar\ell}$. Hence

$$(R^{\tau}_{j\bar k})^{t}_{\ell}\, h^{j\bar k} = (R^{\tau}_{\ell\bar k})^{t}_{j}\, h^{j\bar k} = h^{j\bar k}((R^{\tau}_{k\bar\ell})^{k})^{t}_{j} = \overline{(R^{\tau}_{k\bar\ell})^{k}_{j}}\, h^{t\bar j} = \overline{(R^{\tau}_{j\bar\ell})^{k}_{k}}\, h^{t\bar j} = 0.$$

Thus (9) holds. Similarly

$$(R^{\tau}_{\bar k})^{t}_{j}\, h^{j\bar k} = -h^{j\bar k}((R^{\tau}_{k})^{*})^{t}_{j} = -\overline{(R^{\tau}_{k})^{k}_{j}}\, h^{t\bar j} = 0,$$

proving (10). Since $\overline{\tau_k^{\ell}}\, h^{j\bar k} = h^{k\bar\ell}(\tau^*)^{j}_{k}$, we find that the left side of (11) is equal to

$$(R^{\tau}_{\bar k})^{t}_{\ell}\, h^{j\bar k}(\tau_j^{\ell} + (\tau^*)^{\ell}_{j}) = (R^{\tau}_{\bar k})^{t}_{\ell}\, h^{j\bar k}\, \delta^{\ell}_{j}\, \Pi = (R^{\tau}_{\bar k})^{t}_{j}\, h^{j\bar k} = 0$$

by (9). Hence (11) also holds.

The proof for the fact that $\mu \equiv 0 \mod \Omega, \overline{\Omega}, \Theta$ follows: we calculate $d(d\Phi)$ by (7) and write out the sum of all terms containing $\Omega^{j} \wedge \overline{\Omega^{k}}$ (in terms of the base $\Theta, \Omega, \overline{\Omega}, \Pi, \Phi, \tau, \psi$). Set the sum

$$A_{j\bar k} \wedge \Omega^{j} \wedge \overline{\Omega^{k}}$$

where $A_{j\bar k}$ is a Pfaffian form. Since $d(d\Phi) = 0$, $A_{j\bar k} h^{j\bar k} \equiv 0 \mod \Omega, \overline{\Omega}, \Theta$. By the explicit expression of $A_{j\bar k}$ we find that the property of μ follows. To make the calculation easier to follow, we denote by

$$B_{j\bar k} \wedge \Omega^{j} \wedge \overline{\Omega^{k}} \qquad (\text{resp. } C_{j\bar k} \wedge \Omega^{j} \wedge \overline{\Omega^{k}})$$

the similar term when $d\Phi$ is replaced by $\Phi \quad \tau + \Pi \quad \Phi + \frac{1}{2}\Omega \wedge \psi$ (resp. R^{Φ}). Thus

$$A_{j\bar k} \equiv B_{j\bar k} + C_{j\bar k} \mod \Omega, \overline{\Omega}, \Theta.$$

Clearly

$$B_{j\bar k} \wedge \Omega^{j} \wedge \overline{\Omega^{k}} = (d\Phi)^{(1,1)} \wedge (\tau - \Pi) - \Phi \wedge ((d\tau)^{(1,1)} - (d\Pi)^{(1,1)}) + \frac{1}{2}(d\Omega)^{(1,1)} \wedge \psi - \frac{1}{2}\Omega \wedge (d\psi)''$$

where $('')$ denotes the sum of terms containing $\overline{\Omega^k}$. Therefore

$$B_{j\bar{k}}\wedge\Omega^j\wedge\Omega^{\bar{k}} \equiv \Omega\wedge R^\tau_{\bar{k}}\overline{\Omega^k}\wedge(\tau-\Pi) - \Phi\wedge R^\tau_{j\bar{k}}\Omega^j\wedge\overline{\Omega^k} - \frac{1}{2}\Omega\wedge(i\langle\mu,\Omega\rangle + i\langle\Omega,\mu^{(0,1)}\rangle) \quad \mathrm{mod}\,\Theta.$$

Hence we may set

$$(B_{j\bar{k}})^t \equiv (R^\tau_{\bar{k}})^\ell_j(\tau^t_\ell - \delta^t_\ell\Pi) - R^\tau_{j\bar{k}}\Phi + \frac{i}{2}\delta^t_j h_{\ell\bar{k}}\,\mu^\ell \quad \mathrm{mod}\,\Theta.$$

Therefore, by (9) and (10)

$$(B_{j\bar{k}})^t h^{j\bar{k}} = \frac{i}{2}\mu^t.$$

Now

$$C_{j\bar{k}}\wedge\Omega^j\wedge\overline{\Omega^k} \equiv (d\Omega)''\wedge R^\tau_j\Omega^j + (d\Omega)'\wedge R^\tau_{\bar{k}}\overline{\Omega^k} - \Omega\wedge(d\lambda)'' + i\langle\Omega,\Omega\rangle\wedge\mu \quad \mathrm{mod}\,\Theta.$$

Since $(d\Omega)''=0$, $(d\Omega)'=\Omega\wedge\tau$, and

$$(d\lambda)'' = (dR^\tau_j)^{(0,1)}\wedge\Omega^j + R^\tau_j(d\Omega^j)'' + dR^\tau_{\bar{k}}\wedge\overline{\Omega^k} + R^\tau_{\bar{k}}(d\overline{\Omega^k})'' = (dR^\tau_j)^{(0,1)}\wedge\Omega^j + dR^\tau_{\bar{k}}\wedge\overline{\Omega^k} + R^\tau_{\bar{k}}\overline{\Omega^\ell}\wedge\overline{\tau^k_\ell}$$

$$C_{j\bar{k}}\,\Omega^j\wedge\overline{\Omega^k} \equiv \Omega\wedge\tau\wedge R^\tau_{\bar{k}}\overline{\Omega^k} - \Omega\wedge R^\tau_{\bar{k}}\,\overline{\Omega^\ell}\wedge\overline{\tau^k_\ell} - \Omega\wedge(dR^\tau_j)^{(0,1)}\wedge\Omega^j - \Omega\wedge dR^\tau_{\bar{k}}\wedge\overline{\Omega^k} + i\langle\Omega,\Omega\rangle\wedge\mu$$

Hence

$$(C_{j\bar{k}})^t \equiv -\tau^\ell_j(R^\tau_{\bar{k}})^t_\ell - (R^\tau_{\bar{\ell}})^t_j\,\overline{\tau^\ell_k} - (dR^\tau_{\bar{k}})^t_j + i h_{j\bar{k}}\,\mu^t \quad \mathrm{mod}\,\Omega,\overline{\Omega},\Theta.$$

Therefore, by (10) and (11) we find that

$$(C_{j\bar{k}})^t \equiv i(n-1)\,\mu^t$$

Hence

$$0 \equiv A_{j\bar{k}}\,h^{j\bar{k}} \equiv (B_{j\bar{k}} + C_{j\bar{k}})h^{j\bar{k}} \equiv i(n-\frac{1}{2})\mu \quad \mathrm{mod}\,\Omega,\overline{\Omega},\Theta.$$

Thus we proved that $\mu\equiv 0 \mod \Omega, \overline{\Omega}, \Theta$. In view of (7) we may set

(12) $$\mu = R^\Phi_j\Omega^j + R^\Phi_{\bar{k}}\overline{\Omega^k}$$

where R^Φ_j and $R^\Phi_{\bar{k}}$ are $\mathbb{C}^{n-1}$-valued functions which may depend on ψ. Consider a change

$$\widetilde{\Psi} = \Psi + \Theta F, \qquad F = \overline{F}.$$

By (7) we see that

$$\widetilde{\mu} = \mu + \frac{1}{2}\Omega F$$

Hence

$$i\langle\tilde{\mu}, \Omega\rangle + i\langle\Omega, \tilde{\mu}\rangle = i\langle\mu, \Omega\rangle + i\langle\Omega, \mu\rangle + Fi\langle\Omega, \Omega\rangle$$

Therefore, if we write (Cf. (8))

$$R^\psi \equiv R^\psi_{j\bar{k}}\,\Omega^j\ \overline{\Omega^k} + 2\mathcal{R}R^\psi_{jk}\,\Omega^j\ \Omega^k, \qquad R^\psi_{jk} + R^\psi_{kj} = 0, \quad \mathrm{mod}\ \Theta$$

we find that

$$R^{\tilde{\psi}}_{j\bar{k}} = R^\psi_{j\bar{k}} + Fih_{j\bar{k}}$$

Therefore F is uniquely determined by the condition

$$h^{j\bar{k}}\,R^{\tilde{\psi}}_{j\bar{k}} = 0, \tag{13}$$

i.e.,

$$F = \frac{i}{n-1}R^\psi_{j\bar{k}}\,h^{j\bar{k}}$$

In fact, since $\psi = \overline{\psi}$, $\overline{R^\psi_{j\bar{k}}} = -R^\psi_{k\bar{j}}$. Hence $F = \overline{F}$. We now have a unique $\tilde{\psi}$, which we denote by Ψ. Ψ is globally-defined. By (8) we see that

$$R^\Psi_{j\bar{k}} = ih_{\ell\bar{k}}(R^\Phi_j)^\ell + ih_{j\bar{\ell}}\overline{(R^\Phi_k)^\ell}$$

(14)

$$R^\Psi_{jk} = \frac{i}{2}h_{j\bar{\ell}}\overline{(R^\Phi_{\bar{k}})^\ell} - \frac{i}{2}h_{k\bar{\ell}}\overline{(\Phi^{\Phi}_j)^\ell}$$

Therefore the condition (13) is rewritten as

$$\mathcal{R}(R^\Phi_j)^j = 0 \tag{15}$$

Now we can write

$$d\Psi = \Pi\wedge\Psi + 2i\langle\Phi, \Phi\rangle + R^\Psi$$

(16)

$$R^\Psi = R^\Psi_{j\bar{k}}\,\Omega^j\wedge\overline{\Omega^k} + 2\mathcal{R}R^\Psi_{jk}\,\Omega^j\wedge\Omega^k + \Theta\wedge\gamma$$

where $R^\Psi_{j\bar{k}}$ and R^Ψ_{jk} are determined by (14).

It remains to show that $\gamma \equiv 0 \mod \Omega, \overline{\Omega}, \Theta$. We calculate $d(d\Psi) = 0$ as before. We have mod Θ

$$0 \equiv -\Pi\wedge(R^\Psi)^{(1,1)} + 2i\langle(R^\Phi)^{(1,1)}, \Phi\rangle - 2i\langle\Phi, (R^\Phi)^{(1,1)}\rangle + dR^\Psi_{j\bar{k}}\wedge\Omega^j\wedge\Omega^k + R^\Psi_{\ell\bar{k}}\Omega^j\wedge\tau^\ell_j\wedge\overline{\Omega^k}$$

$$-R^\Psi_{j\bar{\ell}}\,\Omega^j\wedge\overline{\Omega^k}\wedge\overline{\tau^\ell_k} + (dR^\Psi_{jk})^{(0,1)}\wedge\Omega^j\wedge\Omega^k + (d\overline{R^\Psi_{jk}})^{(1,0)}\wedge\overline{\Omega^j}\wedge\overline{\Omega^k} + i\langle\Omega, \Omega\rangle\wedge\gamma$$

Therefore

$$-R^{\Psi}_{j\bar{k}}\Pi + 2i(R^{\tau}_{\bar{k}})^{\ell}_{j} h_{\ell\bar{s}}\Phi^{s} + 2ih_{s\bar{\ell}}\Phi^{s}\overline{(R^{\tau}_{j})^{\ell}_{k}} + dR^{\Psi}_{j\bar{k}} - R^{\Psi}_{\ell\bar{k}}\tau^{\ell}_{j} - R^{\Psi}_{j\bar{\ell}}\overline{\tau^{\ell}_{k}} + ih_{j\bar{k}}\gamma \equiv 0 \quad \mathrm{mod}\, \Omega, \overline{\Omega}, \Theta.$$

Multiplying $h^{j\bar{k}}$ to the above and summing in j, k, we find by (13) and (10)

$$\begin{aligned} i(n-1)\gamma &\equiv R^{\Psi}_{\ell\bar{k}}\tau^{\ell}_{j} h^{j\bar{k}} + R^{\Psi}_{j\bar{\ell}}\overline{\tau^{\ell}_{k}}\, h^{j\bar{k}} \qquad \mathrm{mod}\, \Omega, \overline{\Omega}, \Theta \\ &= R^{\Psi}_{\ell\bar{k}}\tau^{\ell}_{j} h^{j\bar{k}} + R^{\Psi}_{j\bar{\ell}}(\tau^{*})^{j}_{k}\, h^{k\bar{\ell}} \\ &= R^{\Psi}_{\ell\bar{k}}(\tau^{\ell}_{j} + (\tau^{*})^{\ell}_{j})h^{j\bar{k}} \;=\; R^{\Psi}_{j\bar{k}}\delta^{\ell}_{j} h^{j\bar{k}}\Pi \;=\; 0 \end{aligned}$$

Therefore we find that

$$\gamma \equiv 0 \quad \mathrm{mod}\, \Omega, \overline{\Omega}, \Theta.$$

§5. We have explicitly constructed a bundle E over M and a base of Pfaffian forms

$$\Theta, \Omega^{j}, \overline{\Omega^{k}}, \Pi, \tau^{j}_{\ell}, \Phi^{j}, \overline{\Phi^{j}}, \Psi$$

(Θ, Π, Ψ are real) under the condition

$$\tau + \tau^{*} = \Pi I$$

satisfying the following conditions:

(1) the equation $\Theta = \Omega = \overline{\Omega} = 0$ defines the CR structure of M

(2)
$$\begin{aligned} d\Theta &= i\langle\Omega, \Omega\rangle + \Theta\wedge\Pi \\ d\Omega &= \Omega\wedge\tau + \Theta\wedge\Phi \\ d\Pi &= i\langle\Phi, \Omega\rangle + i\langle\Omega, \Phi\rangle + \Theta\wedge\Psi \\ z\,d\tau &= z\tau\wedge\tau + i\langle z, \Omega\rangle\wedge\Phi - i\langle z, \Phi\rangle\wedge\Omega + iz\langle\Omega, \Phi\rangle + \tfrac{1}{2}z\Theta\wedge\Psi + zR^{\tau} \\ d\Phi &= \Phi\wedge\tau + \Pi\wedge\Phi + \tfrac{1}{2}\Omega\wedge\Psi + R^{\Phi} \\ d\Psi &= \Pi\wedge\Psi + 2i\langle\Phi, \Phi\rangle + R^{\Psi} \end{aligned}$$

(3)
$$\begin{aligned} &R^{\tau} = R^{\tau}_{j\bar{k}}\Omega^{j}\wedge\overline{\Omega^{k}} + \Theta\wedge(R^{\tau}_{j}\Omega^{j} + R^{\tau}_{\bar{k}}\overline{\Omega^{k}}) \\ &(R^{\tau}_{j\bar{k}})^{t}_{\ell} = (R^{\tau}_{\ell\bar{k}})^{t}_{j}, \qquad (R^{\tau}_{j\bar{k}})^{*} = R^{\tau}_{k\bar{j}}, \qquad \mathrm{tr}\, R^{\tau}_{j\bar{k}} = 0 \\ &R^{\tau}_{\bar{k}} = -(R^{\tau}_{k})^{*}, \qquad (R^{\tau}_{j})^{j}_{\ell} = 0 \end{aligned}$$

(4)
$$R^{\Phi} = R^{\Phi}_{j\bar{k}}\Omega^{j}\wedge\overline{\Omega^{k}} + R^{\Phi}_{jk}\Omega^{j}\wedge\Omega^{k} + \Theta\wedge(R^{\Phi}_{j}\Omega^{j} + R^{\Phi}_{\bar{k}}\overline{\Omega^{k}}), \qquad R^{\Phi}_{jk} + R^{\Phi}_{kj} = 0$$

$$(R^{\Phi}_{j\bar{k}})^t = (R^{\tau}_{\bar{k}})_j, \qquad (R^{\Phi}_{jk})^t = \frac{1}{2}((R^{\tau}_{k})^t_j - (R^{\tau}_{j})^t_k), \qquad 6\iota(R^{\Phi}_j)^j = 0$$

(5) $$R^{\Psi} = R^{\Psi}_{j\bar{k}}\, \Omega^j \wedge \overline{\Omega^k} + 2\,6\iota\, R^{\Psi}_{jk}\, \Omega^j \wedge \Omega^k + \Theta \wedge (R^{\Psi}_j\, \Omega^j + R^{\Psi}_{\bar{k}}\, \overline{\Omega^k}), \qquad R^{\Psi}_{jk} + R^{\Psi}_{kj} = 0$$

$$R^{\Psi}_{j\bar{k}} = i h_{\ell\bar{k}} (R^{\Phi}_j)^{\ell} + i h_{j\bar{\ell}} \overline{(R^{\Phi}_k)^{\ell}}, \qquad h^{j\bar{k}} R^{\Psi}_{j\bar{k}} = 0$$

$$R^{\Psi}_{jk} = \frac{i}{2} h_{j\bar{\ell}} \overline{(R^{\Phi}_{\bar{k}})^{\ell}} - \frac{i}{2} h_{k\bar{\ell}} \overline{(R^{\Phi}_j)^{\ell}}$$

For a given CR structure over M (with non-degenerate Levi-form $(h_{j\bar{k}})$), E and Pfaffian forms we constructed are locally unique up to isomorphism. In fact we have the following:

PROPOSITION. Given a manifold $\tilde{E}$ with a projection $\tilde{\rho}: \tilde{E} \longrightarrow M$. Assume that we are given Pfaffian forms

$$\tilde{\Theta}, \tilde{\Omega} = (\tilde{\Omega}^1, \ldots, \tilde{\Omega}^{n-1}), \tilde{\Pi}$$

where $\tilde{\Theta}$ and $\tilde{\Pi}$ are real, such that for Pfaffian forms $\tilde{\tau} = (\tilde{\tau}^{\ell}_k)$, $\tilde{\varphi} = (\tilde{\varphi}^1, \ldots, \tilde{\varphi}^{n-1})$, real $\tilde{\psi}$, satisfy:

$$d\tilde{\Theta} = i\langle \tilde{\Omega}, \tilde{\Omega} \rangle + \tilde{\Theta} \wedge \tilde{\Pi}$$

$$d\tilde{\Omega} = \tilde{\Omega} \wedge \tilde{\tau} + \tilde{\Theta} \wedge \tilde{\varphi}$$

$$d\tilde{\Pi} = i\langle \tilde{\varphi}, \tilde{\Omega} \rangle + i\langle \tilde{\Omega}, \tilde{\varphi} \rangle + \tilde{\Theta} \wedge \tilde{\psi}.$$

Moreover, the above Pfaffian forms, together with their bar of non-real elements, form a base under the condition

$$\tilde{\tau} + \tilde{\tau}^* = \Pi I.$$

Assume further that $\tilde{\Theta} = \tilde{\Omega} = 0$ defines the given CR structure.

Pick $\tilde{F} \in \tilde{E}$. Then 1) there is an open neighborhood $\tilde{E}_1$ of $\tilde{F}$ and a fiber preserving map

$$\iota: \tilde{E}_1 \longrightarrow E$$

such that

$$\iota^*\Theta = \tilde{\Theta}, \qquad \iota^*\Omega = \tilde{\Omega}, \qquad \iota^*\Pi = \tilde{\Pi}$$

2) We can find unique $\tilde{\tau}, \tilde{\Phi}, \tilde{\Psi}$ on $\tilde{E}$ such that $\tilde{\Theta}, \tilde{\Omega}, \tilde{\Pi}, \tilde{\tau}, \tilde{\Phi}, \tilde{\Psi}$ are normal Cartan connection forms. Moreover, $\iota^*\tau = \tilde{\tau}$, $\iota^*\Phi = \Phi$ and $\iota^*\Psi = \Psi^*$ where ι is the map given in 1) above.

PROOF. Since $\tilde{\Theta} = \tilde{\Omega} = 0$ defines our CR structure, and since real annihilators of T' form a 1-dimensional vector space at each point in M, we have locally

(6) $$\tilde{\Theta} = \tilde{p}\, \rho^* \theta_0$$

where $\tilde{p}$ is a C^∞-function on a neighborhood of $\tilde{F}$. (Here we assumed ρF is the neighborhood U_0, Cf: (0.1), (0.2) in §2. Moreover

$$\tilde{\Omega} = (\rho^* \omega_0)\tilde{t} + \Theta\tilde{v} \tag{7}$$

where $\tilde{t} = (\tilde{t}_k^\ell)$, non-singular and $\tilde{v} = (\ldots, \tilde{v}^j, \ldots)$ are also C^∞- on a neighborhood of $\tilde{F}$. Then

$$\begin{aligned} d\tilde{\Theta} &= i\tilde{p}\rho^*\langle\omega_0, \omega_0\rangle + \rho^*\theta_0 \wedge(-d\tilde{p} + \tilde{p}\rho^*\pi_0) \\ &= i\tilde{p}\langle(\tilde{\Omega} - \tilde{\Theta}\tilde{v})\tilde{t}^{-1}, (\tilde{\Omega} - \tilde{\Theta}\tilde{v})\tilde{t}^{-1}\rangle + \tilde{\Theta}\wedge(-d\log\tilde{p} + \rho^*\pi_0) \\ &= i\langle\tilde{\Omega}\,\tilde{p}\tilde{t}^{-1}(\tilde{t}^{-1})^*, \tilde{\Omega}\rangle + \tilde{\Theta}\wedge(-d\log\tilde{p} + \rho^*\pi_0 - i\langle\tilde{v}\,\tilde{p}\tilde{t}^{-1}(\tilde{t}^{-1})^*, \tilde{\Omega}\rangle + i\langle\tilde{\Omega}\tilde{p}\tilde{t}^{-1}(\tilde{t}^{-1})^*, \tilde{v}\rangle \end{aligned}$$

Hence by the condition of linear independence, we see that

$$\tilde{p}I = \tilde{t}^*\tilde{t} \tag{8}$$

$$\tilde{\Pi} \equiv -d\log\tilde{p} + \rho^*\pi_0 + i\langle\tilde{\Omega}, \tilde{v}\rangle - i\langle\tilde{v}, \tilde{\Omega}\rangle \quad \mathrm{mod}\,\tilde{\Theta}. \tag{9}$$

Since $\tilde{\Pi}$ is non-zero everywhere $\mathrm{mod}\,\tilde{\Omega}, \overline{\tilde{\Omega}}, \tilde{\Theta}$, we find that $d\tilde{p} \neq 0$ everywhere. Hence, if we restrict to a small neighborhood of $\tilde{F}$, there is a bundle $\tilde{E}_1 \longrightarrow M$ with a projection $\tilde{\rho}_1$ such that $\tilde{p}$ is its fiber coordinate. Since our expression of $d\tilde{\Theta}$ shows that $(h_{j\bar{k}})$ is Levi-form with respect to $\tilde{\Theta}, \tilde{\Omega}, \tilde{p} > 0$. Since the fiber chart p in E_1 (Cf. (1)′ §2) has the range $\mathbb{R}^+$, it is now clear that there is a C^∞ injection

$$\iota_1 : \tilde{E}_1 \longrightarrow E_1$$

such that $\iota_1^*\Theta = \tilde{\Theta}$ (cf. (1), §2). The rest of the proof goes similarly. Namely, by (7) and §2 (0.1) & (0.2)

$$\begin{aligned} d\tilde{\Omega} &= -(\rho^*\omega_0)\wedge d\tilde{t} + \rho^*\omega_0\wedge\tau_0\tilde{t} + \theta_0\wedge\varphi_0\tilde{t} - \Theta\wedge d\tilde{v} + (i\langle\tilde{\Omega}, \tilde{\Omega}\rangle + \tilde{\Theta}\wedge\tilde{\Pi})\tilde{v} \\ &= -(\tilde{\Omega} - \tilde{v}\tilde{\Theta})\wedge(\tilde{t}^{-1}d\tilde{t} + \tilde{t}^{-1}\tau_0\tilde{t}) + i\langle\tilde{\Omega}, \tilde{\Omega}\rangle\tilde{v} + \tilde{\Theta}\wedge(-d\tilde{v} + \tilde{p}^{-1}\varphi_0\tilde{t} + \tilde{\Pi}\tilde{v}) \\ &= \tilde{\Omega}\wedge(-\tilde{t}^{-1}d\tilde{t} + \tilde{t}^{-1}\tau_0\tilde{t}) + i\langle\tilde{\Omega}, \tilde{\Omega}\rangle\tilde{v} + \tilde{\Theta}\wedge(-d\tilde{v} + \tilde{v}\tilde{t}^{-1}d\tilde{t} + \tilde{v}\tilde{t}^{-1}\tau_0\tilde{t} + \tilde{p}^{-1}\varphi_0\tilde{t} + \tilde{\Pi}\tilde{v}) \end{aligned}$$

Therefore

$$\begin{aligned} \tilde{\tau} &\equiv -\tilde{t}^{-1}d\tilde{t} \quad \mathrm{mod}(\tilde{\Omega}, \overline{\tilde{\Omega}}, \tilde{\Theta}) \\ \tilde{\varphi} &\equiv -d\tilde{v} + \tilde{v}\tilde{t}^{-1}d\tilde{t} + \tilde{\Pi}\tilde{v} \quad \mathrm{mod}(\tilde{\Omega}, \overline{\tilde{\Omega}}, \tilde{\Theta}). \end{aligned}$$

Now it is clear from the construction of the bundle E_2 in §2 that our $\tilde{\Theta}, \tilde{\Omega}, \tilde{\varphi}$ is a copy of Θ, Ω, φ restricted to an open subset. In view of (9) there is a C^∞-function $\tilde{S}$ such that

$$\tilde{\Pi} = -d\log\tilde{p} + \rho^*\pi_0 + i\langle\tilde{\Omega}, \tilde{v}\rangle - i\langle\tilde{v}, \tilde{\Omega}\rangle + \tilde{S}.\tilde{\Theta}$$

By calculating $d\tilde{\Pi}$ we find similarly that $d\tilde{S} \neq 0$ everywhere. Then our contention 1) is again clear from the construction of the bundle E in §2. Our contention 2) is the consequence of our whole discussion in §3 and §4, q.e.d.

§6. Quadrics.

We consider a very important example of a CR structure. This one plays the role of the euclidean space in the analogy with the Riemann geometry.

We denote the general element in $\mathbb{C}^n$ by

$$(z, w), \qquad z = (z^1, \ldots, z^{n-1}) \in \mathbb{C}^{n-1}, \qquad w \in \mathbb{C}$$

In $\mathbb{C}^n$ we consider a real hypersurface Q given by the equation

(1) $$Q\colon y = \frac{1}{2}\langle z, z\rangle,$$

where $w = x + iy$, $(x, y \in \mathbb{R})$.

We regard Q endowed with the CR-structure induced by the ambiant space $\mathbb{C}^n$. Let

(2) $$\iota\colon Q \longrightarrow \mathbb{C}^n$$

be the injection. The defining equation T'' of Q, which we denote by T''_Q, is the annihilator of ι^*dh, where h runs through germs of holomorphic functions of $\mathbb{C}^n$ at points in Q. Therefore T''_Q is defined by the equation

$$\iota^*dw = dx + \frac{i}{2}\langle dz, z\rangle + \frac{i}{2}\langle z, dz\rangle = 0$$

$$dz^j = 0$$

where we regarded (z, x) as a global chart of Q. Since we want one of them to be real as in §2, we subtract $i\langle dz, z\rangle$ from the first. Therefore T''_Q is defined by the equation

(3) $$\theta_Q = dx - \frac{i}{2}\langle dz, z\rangle + \frac{i}{2}\langle z, dz\rangle = 0$$
$$\omega_Q = dz$$

Clearly

(4) $$d\theta_Q = i\langle \omega_Q, \omega_Q\rangle$$
$$d\omega_Q = 0$$

Thus this is the case $\pi_0 = \tau_0 = \varphi_0 = 0$ in the formula (0) §2.

We next show that Q is essentially a homogeneous space. In fact, the CR structure of Q extends to the one-point compactification of Q, and this compactification is a homogeneous space. To see this, we denote the general element of $\mathbb{C}^{n+1}$ by

$$\xi = (\xi^0, \dots, \xi^n)$$

and we consider the complex projective space $\mathbf{P}^n$ whose general element is given by the homogeneous coordinate

$$[\xi] = [\xi^0, \dots, \xi^n]$$

We consider the embedding

$$\mathbb{C}^n \ni (z, w) \longrightarrow [1, z, w] \in \mathbf{P}^n$$

and identify $\mathbb{C}^n$ as the open subset of $\mathbf{P}^n$. Then Q is the intersection with $\mathbb{C}^n$ of a real hypersurface $\widetilde{Q}$ in $\mathbf{P}^n$ defined by

$$\frac{1}{2i}\left(\frac{\xi^n}{\xi^0} - \frac{\overline{\xi^n}}{\overline{\xi^0}}\right) = \frac{1}{2} h_{j\bar{k}}\, \xi^j \overline{\xi^k} / |\xi^0|^2,$$

i.e.,

$$\widetilde{Q}\colon \xi^0 \overline{\xi^n} - \overline{\xi^0}\, \xi^n + h_{j\bar{k}}\, \xi^j \overline{\xi^k} = 0$$

Then it is easy to check

(5) **PROPOSITION** $\widetilde{Q} = Q \cup P_\infty$, $P_\infty = \{[\xi];\ \xi \neq 0,\ \xi^0 = h_{j\bar{k}}\, \xi^j \xi^{\bar{k}} = 0\}$

To see that a Lie group of CR isomorphism of $\widetilde{Q}$ acts transitively on $\widetilde{Q}$ we introduce the following:

$$\xi = (\xi^0, \xi', \xi^n), \qquad \xi' = (\xi', \dots, \xi^{n-1}), \qquad \xi'' = (\xi^0, \xi^n)$$

(6)

$$\alpha \wedge \overline{\beta} = \alpha^0 \overline{\beta^n} - \alpha^n \overline{\beta^0} \qquad \text{for } \alpha = (\alpha^0, \alpha^n) \text{ and } \beta = (\beta^0, \beta^n)$$

$$(\xi, \xi) = \frac{1}{\iota} \xi'' \ \overline{\xi^n} + \langle \xi', \xi' \rangle$$

Now the equation of $\widetilde{Q}$ is given by

(7) $$(\xi, \xi) = 0$$

If $(h_{j\bar{k}})$ has p positive eigenvalues and q negative eigenvalues, (ξ, ξ) has $p+1$ positive eigenvalues and $q+1$ negative eigenvalues. We set

(8) $$G^\# = \{g \in SL(n+1, \mathbb{C}^n);\ (g\xi, g\xi) = (\xi, \xi)\}$$

Clearly $G^\#$ is a Lie group isomorphic to $SU(p+1, q+1)$ and acts as an isomorphism group of the CR structure on $\widetilde{Q}$. For $g \in G^\#$, denote by $[g]$ the map induced on $\widetilde{Q}$. Since $\widetilde{Q}$ is real codimension 1 and the induced map of g on $\mathbf{P}^n$ is holomorphic, it is clear that $[g]$ is the identity map if and only if g is a scalar matrix. Since $\det g = 1$, it then follows that such g is equal to λI with $\lambda^{n+1} = 1$. Hence we have the exact sequence

$$0 \longrightarrow \mathbb{Z}_{n+1} \longrightarrow G^{\#} \longrightarrow G \longrightarrow \{1\}$$

(9)

$$\mathbb{Z}_{n+1} = \{\lambda I;\ \lambda^{n+1} = 1\},\quad G = \{[g];\ g \in G^{\#}\}$$

Clearly G is a Lie group which acts effectively on $\widetilde{Q}$. It is a group of CR isomorphisms of $\widetilde{Q}$. To show that G acts transitively on $\widetilde{Q}$, we consider for each $(z, x) \in Q$ a linear map $T_{(z,x)}: \xi \longrightarrow \xi_*$ of $\mathbb{C}^{n+1}$ given by

$$\xi_*^0 = \xi^0$$

(10)
$$\xi_*' = \xi' + z\xi^0$$

$$\xi_*^n = \xi^n + \xi^0 w + i\langle \xi', z\rangle, \qquad w = x + \frac{i}{2}\langle z, z\rangle$$

We see easily that the expression of (ξ, ξ) in (6) that $T_{(z,x)} \in G^{\#}$.
The matrix representation is given by

(11)
$$T_{(z,x)}:\ (\xi^0, \ldots, \xi^n)\begin{pmatrix} 1 & z & w \\ 0 & I & iz^* \\ 0 & 0 & 1 \end{pmatrix} = (\xi_*^0, \ldots, \xi_*^n)$$

where z^* denotes the column vector whose j-th component is equal to $h_{j\bar{k}}\,\overline{z^k}$. We calculate easily that

$$T_{(z,x)}T_{(z_1,x_1)} = T_{(z+z_1,\ x+x_1+6\,i\langle z_1, z\rangle)}$$

(12)

$$T_{(z,x)}^{-1} = T_{(-z,-x)}$$

Hence $S^{\#} = \{T_{(z,x)}:\ (z, x) \in Q\}$ is a subgroup of $G^{\#}$. Clearly

(13)
$$T_{(z,x)}(1, 0, \ldots, 0) = (1, z, x)$$

We set $S = \{[T_{(z,x)}]:\ (z, x) \in Q\}$. Clearly the natural map $S^{\#} \longrightarrow S$ is an isomorphism. Thus Q may be considered as a group. In this sense, Q is called *Heisenberg group.*

(14) **PROPOSITION.** G *acts transitively on* $\widetilde{Q}$.

PROOF. By (13) we find that for any two points in Q there is an element of G which sends one to another. Since (ξ, ξ) is invariant under the interchange of (ξ^0, ξ^n) to $(\xi^n, -\xi^0)$, the same holds for $Q' = [\xi] \in \widetilde{Q};\ \xi^n \neq 0$. Since $Q \cap Q' \neq \emptyset$, it suffices to check that any element in $\widetilde{Q} - (Q \cup Q')$ can be sent into $Q \cup Q'$. This can be easily done by a suitable $T_{(2,x)}$.

We denote by H the isotropy group of $[1, 0, \ldots, 0]$ in G. Set

$$H^* = \{g \in G^{\#};\ [g] \in H\}$$

By calculation we find that $H^{\#}$ consists of the following transformations:
For $a, b, \in \mathbb{C}$, $\beta \in \mathbb{C}^{n-1}$, $\gamma = (\gamma^j_k)$ $(i, k = 1, \ldots, n-1)$ such that

$$(15) \qquad a \neq 0, \qquad \gamma\gamma^* = I, \qquad \frac{1}{2}\langle \beta, \beta \rangle = -\mathrm{Im}\,\frac{b}{a}, \qquad \frac{\bar{a}}{a} \det \gamma = 1$$

we define

$$T_{(a, b, \beta, \gamma)}: \xi \longrightarrow \xi_*$$

by

$$(16) \qquad \begin{aligned} \xi^0_* &= a\xi^0 + \langle \xi', \mu \rangle + b\xi^n, \qquad \mu = i\bar{a}\beta\gamma^* \\ \xi'_* &= \xi'\gamma + \beta\xi^n \\ \xi^n_* &= \frac{1}{\bar{a}}\xi^n \end{aligned}$$

The matrix representation is given by

$$(17) \qquad (\xi^0, \ldots, \xi^n) \begin{pmatrix} a & 0 & 0 \\ \mu^* & \gamma & 0 \\ b & \beta & (\bar{a})^{-1} \end{pmatrix} = (\xi^0_*, \ldots, \xi^n_*)$$

By calculation we find that

$$T_{(a, b, \beta, \gamma)} T_{(a', b', \beta', \gamma')} = T_{(\tilde{a}, \tilde{b}, \tilde{\beta}, \tilde{\gamma})}, \quad \text{where}$$

$$(18) \qquad \begin{aligned} \tilde{a} &= a'a \\ \tilde{b} &= b'a + (\bar{a}')^{-1}b - ia\langle \beta'\gamma, \beta \rangle \\ \tilde{\beta} &= \beta'\gamma + (\bar{a}')^{-1}b \\ \tilde{\gamma} &= \gamma'\gamma \end{aligned}$$

$$T^{-1}_{(a, b, \beta, \gamma)} = T_{(a', b', \beta', \gamma')}, \quad \text{where}$$

$$(19) \qquad \begin{aligned} a' &= a^{-1}, \qquad b' = -b\bar{a}a^{-1} - i\bar{a}\langle \beta, \beta \rangle \\ \beta' &= -\bar{a}\beta\gamma^{-1}, \quad \gamma' = \gamma^{-1} \end{aligned}$$

By the definition of H, we have the canonical map

(20) $$\rho_G\colon G \longrightarrow G/H = \tilde{Q} \quad (g \longmapsto gH)$$

We set

(21) $$B = \rho_G^{-1}(Q), \quad \rho_B = \rho_G|B\colon B \longmapsto Q$$

B is an open subset of G. Since H is the isotropy group of [1, 0, ..., 0] and $T_{(z,x)} \longrightarrow [1, z, x]$ is bijective, we see that we have a diffeomorphism

(22) $$S\times H \ni T_{(z,x)}\times T_{(a,b,\beta,\gamma)} \longmapsto T_{(z,x)}\circ T_{(a,b,\beta,\gamma)} \in B.$$

The above induces a chart (with restraint)

(23) $$(z, x, a, b, \beta, \gamma) \in B$$

where a, b, β, γ are restrained by the condition (15) and $T_{(a,b,\beta,\gamma)}$ is considered mod $\mathbb{Z}_{n+1}$ (Cf. (9)). Clearly

$$\rho_G(z, x, a, b, \beta, \gamma) = (z, x)$$

We may regard B as a principal bundle over Q with structure group H. B is also an open subset of the Lie group G. We now look at the left-invariant Pfaffian forms (i.e., Mauer-Cartan forms) of G restricted to B. We denote by Θ_Q (rest. Ω_Q^j) the one which agrees at the identity element with $d(x\circ\rho_B)$ (rest. with $d(z^j\circ\rho_B)$). We are going to find the expression of Θ_Q, Ω_Q in terms of the chart given in (23). We fix first $g_1 = (z_1, x_1, a_1, b_1, \beta_1, \gamma_1) \in B$ and try to find $(\Theta_Q)_{g_1}$, $(\Omega_Q)_{g_1}$. We then define a map given by

(24) $$\mathcal{L}\colon g = (z, x, a, b, \beta, \gamma) \longrightarrow g_1^{-1}g = (z_*, x_*, a_*, b_*, \beta_*, \gamma_*)$$

defined for g in a neighborhood of g_1. Then, by the left-invariance we see that

(25) $$(\Theta_Q)_{g_1} = \mathcal{L}^*(dx)_I, \qquad (\Omega_Q)_{g_1} = \mathcal{L}^*(dz)_I$$

where I denotes the identity element in G. Therefore when we regard x_* as a function of variables $(z, x, a, b, \beta, \gamma)$ and consider dx_*, $(\Theta_Q)_{g_1}$ is given by evaluating the differential form dz_* at $(z_1, x_1, a_1, b_1, \beta_1, \gamma_1)$. This process we may express symbolically by

(26) $$(\Theta_Q)_{g_1} = (dx_*)_{g=g_1}$$

Similarly for Ω_Q. Thus our task is to find the expression of x_*, z_* as functions of $(z, x, a, b, \beta, \gamma)$. By (24) we see that they are determined by the equation

(27) $$T_{(z_*, x_*)} T_{(a_*, b_*, \beta_*, \gamma_*)}$$

$$= T^{-1}_{(a_1, b_1, \beta_1, \gamma_1)} T_{(z-z_1, x-x_1-\Re i\langle z, z_1\rangle)} T_{(a, b, \beta, \gamma)}$$

(Cf. (12)). When we apply the above to (1, 0, ..., 0), we find by the definition of T's that the left-hand side of the above is equal to $(a_*, a_* z_*, a_* w_*)$. Therefore we find x_*, z_* by calculating the right-hand side applied to (1, 0, ..., 0). We thus find that

$$a_* = \frac{a}{a_1} + ia\langle z-z_1, \beta_1\rangle - a(w-\overline{w_1} - i\langle z, z_1\rangle)(b_1 \frac{\overline{a_1}}{a_1} + i\overline{a_1}\langle \beta_1, \beta_1\rangle)$$

(28) $$z_* = \frac{1}{a^*}(a(z-z_1) - a\overline{a_1}\beta_1(w-\overline{w_1} - i\langle z, z_1\rangle))\gamma_1^{-1}$$

$$x_* = \frac{1}{a_*}a\overline{a_1}(w-\overline{w_1} - i\langle z, z_1\rangle)$$

Noting that $(a_*)_{g=g_1} = 1$, we see easily

$$(dx_*)_{g=g_1} = |a_1|^2(dx - \frac{i}{2}\langle dz, z_1\rangle + \frac{i}{2}\langle z_1, dz\rangle)_{g_1}$$

$$(dz_*)_{g=g_1} = (a_1(dz)_{g_1} - \beta_1(dx_*)_{g=g_1})\gamma_1^{-1}$$

Therefore we find that in terms of the chart $(z, x, a, b, \beta, \gamma)$

$$\Theta_Q = |a|^2\theta_Q$$

(29)

$$\Omega_Q = (a\omega_Q - \beta\Theta_Q)\gamma^{-1}$$

(Cf. (3)). Comparing the above with (1) and (4) in §2, we find that Mauer-Cartan forms on B and the Cartan-connection forms must be related. Actually under a suitable identification of B with the bundle E over Q we constructed previously, Cartan connection forms of Q form a base of the Mauer-Cartan forms. We discuss this in the next section.

§7. Quadrics (continued)

By (29) §6 we see that

(1) $$(d\Theta_Q)_I = i\langle (dz)_I, (dz)_I\rangle + (dx)_I \wedge (-d|a|^2)_I$$

because $a = 1$ when $g = I$. In view of (7) §2 the above suggests that we denote by Π_Q the left-invariant form on B determined by the condition

(2) $$(\Pi_Q)_I = -(d|a|^2)_I$$

Then because of the left-invariance it follows by (1)

(3) $$d\Theta_Q = i\langle\Omega_Q, \Omega_Q\rangle + \Theta_Q \wedge \Pi_Q$$

As in the case of Θ_Q, for any fixed $g_1 = (z_1, x_1, a_1, b_1, \beta_1, \gamma_1)$

$$(\Pi_Q)_{g_1} = -(d|a_*|^2)_{g=g_1}$$

where a_* is the function in the variable $g = (z_*, x_*, a, b, \beta, \gamma)$ given in (28) §2. The formula for a_* shows that

$$(da_*)_{g=g_1} = (\frac{da}{a_1})_{g_1} + ia_1\langle dx, \beta_1\rangle_{g_1} - a_1(dx - 6i\langle dz, z_1\rangle)_{g_1}(b_1\frac{\overline{a_1}}{a_1} + i\overline{a_1}\langle\beta_1, \beta_1\rangle)$$

$$= (d\log a)_{g_1} + i\langle\Omega_Q\gamma + \beta\Theta_Q, \beta\rangle_{g_1} - (\Theta_Q)_{g_1}(\frac{b_1}{a_1} + i\langle\beta_1, \beta_1\rangle)$$

$$= (d\log a)_{g_1} + i\langle\Omega_Q, \beta\gamma^{-1}\rangle_{g_1} - (\frac{b}{a}\Theta_Q)_{g_1}.$$

Since $a_* = 1$ when $g = g_1$, $(d|a_*|^2)_{g=g_1} = (da_*)_{g=g_1} + (d\overline{a_*})_{g=g_1}$. Therefore we see that in terms of the chart $(z, x, a, b, \beta, \gamma)$

(4) $$\Pi_Q = -d\log|a|^2 - i\langle\Omega_Q, \beta\gamma^{-1}\rangle + i\langle\beta\gamma^{-1}, \Omega_Q\rangle + (\frac{b}{a} + \frac{\overline{b}}{\overline{a}})\Theta_Q$$

Compare the expression of Θ_Q, Ω_Q, and Π_Q with that of Θ, Ω, Π given in (1), (4), and (6) in §2. We see that there is a map of B to E which sends Θ, Ω, Π to $\Theta_Q, \Omega_Q, \Pi_Q$, respectively. Namely, we denote here by E the bundle of Cartan connection over Q. E has a chart (z, x, t, v. s) once we fix Pfaffian forms θ_0, ω^j_0, π_0, $(\tau_0)^j_k$, φ^j_0 satisfying the conditions in (0) §2. In Q we may set

(5) $$\theta_0 = \theta_Q, \qquad \omega_0 = \omega_Q, \qquad \pi_0 = \tau_0 = \varphi_0 = 0.$$

Thus (z, x, p, t, v, s) is a global chart of E. We then define a map

(6) $$\tilde{\iota}: B \longrightarrow E$$

over Q given by the equation

(7) $$\begin{aligned} p &= |a|^2 \\ t &= a\gamma^{-1} \\ v &= -\beta\gamma^{-1} \\ s &= 2 6i\frac{b}{a} \end{aligned}$$

To see that the above definition makes sense we have to check that the above equation preserves the restraint conditions of our charts. We recall that they are

(8) $$p > 0, \qquad t^*t = pI, \qquad s = \overline{s},$$

$$a \neq 0, \qquad \gamma\gamma^* = I, \qquad \frac{1}{2}\langle\beta,\beta\rangle = -\mathrm{Im}\frac{b}{a} \qquad \frac{a}{\overline{a}}\det\gamma = 1,$$

(9)

$$T_{(a,b,\beta,\gamma)} \text{ is considered modulo } \mathbb{Z}_{n+1}$$

We easily see that (9) imply (8) under the equation (7). Hence $\widetilde{\iota}$ is a C^∞ bundle map and

(10) $$\widetilde{\iota}^*\Theta = \Theta_Q, \qquad \widetilde{\iota}^*\Omega = \Omega_Q, \qquad \widetilde{\iota}^*\Pi = \Pi_Q.$$

(11) **PROPOSITION.** *$\widetilde{\iota}$ is a diffeomorphism of the bundle* B *to the bundle* E *over* Q.

PROOF. We show that for each p, t, v, s satisfying (8) there is unique solution a, b, β, γ of the equation (7) satisfying (9). Obviously $a = \sqrt{p}e^{i\theta}$, where θ is yet to be determined modulo (n+1). Then $\gamma = \sqrt{p}e^{i\theta}t^{-1}$. Then it follows that $\gamma\gamma^* = 1$. Now the condition $(a/\overline{a})\det\gamma = 1$ determines θ modulo (n+1). $\beta = -v\gamma$. Then $\mathrm{Im}(b/a) = -\langle\beta,\beta\rangle/2 = -\langle v,v\rangle/2$. Therefore $b = a(s - i\langle v,v\rangle)/2$. Thus the unique solution is given by

$$a = \sqrt{p}e^{i\theta}, \qquad e^{i(n-1)\theta} = p^{-\frac{1}{2}(n-1)}\det t$$

$$b = \sqrt{p}e^{i\theta}(s - i\langle v,v\rangle)/2$$

(12) $$\gamma = \sqrt{p}e^{i\theta}t^{-1}$$

$$\beta = -\sqrt{p}e^{i\theta}vt^{-1}$$

Thus $\widetilde{\iota}^{-1}$ is also C^∞, q.e.d.

Now we claim that under the identification by $\widetilde{\iota}$, all the Cartan connection forms of Q are the left-invariant Pfaffian forms on B. Namely, for any fixed $g_1 \in B \subset G$ we consider a map $L_{g_1}: g \longrightarrow g_1^{-1}g$ of an open neighborhood U_1 of g_1 to an open neighborhood of I. Then the structure equation of $\Theta, \Omega, \Pi, \tau, \Phi, \Psi$ restricted to U is mapped by $(L_{g_1})^*$ to the structure equation of $\Theta, \Omega, \Pi, L_{g_1}^*\tau$, $L_{g_1}^*\Phi$, $L_{g_1}^*\Psi$ (cf. (10)) restricted to U_1. The curvature terms of the latter are $L_{g_1}^*R^\tau, L_{g_1}^*R^\Phi, L_{g_1}^*R^\Psi$. Hence they also satisfy the symmetry and trace conditions. Therefore, by the uniqueness theorem (cf. §5) it follows that $\tau = L_{g_1}^*\tau$, $\Phi = L_{g_1}^*\Phi$, $\Psi = L_{g_1}^*\Psi$. Thus our claim is proved.

We next identify τ, Φ, Ψ as Mauer-Cartan forms and determine the curvature terms of Q. By (29) §6 and (4) we find that

$$(d\Omega_Q)_I = -(dz)_I \wedge (d\gamma^{-1})_I + (da)_I \wedge (dz)_I + (dx)_I \wedge (d\beta)_I$$

$$= (\Omega_Q)_I \wedge (-d(\gamma^{-1}) - da)_I + (\Theta_Q)_I \wedge (d\beta)_I$$

$$(d\Pi_Q)_I = i\langle\Omega_Q, d\beta\rangle_I + i\langle d\beta, \Omega_Q\rangle_I - (\Theta_Q)_I \wedge (2d(\mathcal{R}\frac{b}{a}))_I$$

The formula (14) §2 suggests that we denote by τ_Q, Φ_Q, and Ψ_Q the left-invariant Pfaffian forms on B determined by the conditions:

$$(13) \qquad \begin{aligned} (\tau_Q)_I &= -(d(\gamma^{-1}) + da)_I = (d\gamma - da)_I \\ (\Phi_Q)_I &= (d\beta)_I \\ (\Psi_Q)_I &= -2(d\,\mathcal{R}\frac{b}{a})_I = -(db + d\bar{b})_I = -2(db)_I \end{aligned}$$

(Cf. (9)). Then by the left-invariance it follows that

$$(14) \qquad \begin{aligned} d\Omega_Q &= \Omega_Q \wedge \tau_Q + \Theta_Q \wedge \Phi_Q \\ d\Pi_Q &= i\langle \Omega_Q, \Phi_Q \rangle + i\langle \Phi_Q, \Omega_Q \rangle + \Theta_Q \wedge \Psi_Q \end{aligned}$$

By the definition of τ_Q and (9)

$$\begin{aligned} (\tau_Q + \tau_Q^*)_I &= (-d\gamma^{-1} - d\gamma^{-1*} - da - d\bar{a})_I \\ &= -(da + d\bar{a})_I = -(d|a|^2)_I = (\Pi_Q)_I , \end{aligned}$$

Therefore

$$(15) \qquad \tau_Q + \tau_Q^* = \Pi_Q$$

where the scalar-valued form Π_Q is considered as scalar matrix-valued. To find their chart expression of τ_Q, Φ_Q, Ψ_Q, we need the expression of $a_*, \gamma_*, \beta_*, b_*$ in terms of $(z, x, a, b, \beta, \gamma)$ given by (27) §6. Since we already have a_*, z_*, x_* in (28) §6, γ_*, β_*, b_* can be obtained by the formula:

$$(16) \qquad T_{(a_*, b_*, \beta_*, \gamma_*)} = T_{(-z_*, -x_*)} T^{-1}_{(a_1, b_1, \beta_1, \gamma_1)} T_{(z - z_1, x - x_1 - \mathcal{R} i\langle z, z_1 \rangle)} T_{(a, b, \beta, \gamma)}$$

When we apply the left-hand side of the above to $(0, ..., 0, 1)$ and $(0, \gamma', 0)$, we find by (16) (or (17)) in §6 that we obtain

$$(b_*, \beta_*, (\bar{a}_*)^{-1}) \qquad \text{and} \qquad (\langle \xi', \mu_* \rangle, \xi' \gamma_*, 0),$$

respectively. Therefore γ_*, β_*, b_* is obtained by applying the right-hand side to $(0, ..., 0, 1)$ and $(0, \gamma', 0)$. By means of (10) and (16) in §6 we thus find that

(17)

$$b_* \equiv \left(\frac{b}{a_1} - \frac{\bar{a}_1}{\bar{a}}\frac{b_1}{a_1}\right) + i\left(\langle \beta, \beta_1 \rangle - \frac{\bar{a}_1}{a}\langle \beta_1, \beta_1 \rangle\right) + b\left(\frac{a_*}{a} - \frac{1}{a_1}\right) - i\left(\frac{b_1}{a_1} + i\langle \beta_1, \beta_1 \rangle\right)\langle \beta, a_1(z - z_1) \rangle \mod m^2_{g_1}$$

(18) $$\beta_* \equiv (\beta - \frac{\overline{a}_1}{\overline{a}}\beta_1)\gamma_1^{-1} + \frac{b}{a}z_* - i\langle\beta, a_1(z-z_1)\rangle\beta_1\gamma_1^{-1} \mod m_{g_1}^2$$

(19) $$u\gamma_* \equiv (u\gamma - ia(z-z_1)\langle u\gamma, \beta\rangle)\gamma_1^{-1} + i\langle u\gamma, \beta\rangle\overline{a}_1\, a(w - \overline{w}_1 - i\langle z_1, z\rangle)\beta_1\gamma_1^{-1}$$
$$- i\langle u\gamma, a_1(z-z_1)\rangle\beta_1\gamma_1^{-1} \mod m_{g_1}^2$$

where $u = (u^1, \ldots, u^{n-1})$ denotes an independent variable in $\mathbb{C}^{n-1}$ and m_{g_1} denotes the ideal of g_1 on E. Then, as we obtained Θ_Q, Ω_Q, and Π_Q we find that

(20)

$$u\tau_Q = u(d\gamma)\gamma^{-1} - u\,d\log a - i\langle u, \beta\gamma^{-1}\rangle\Omega_Q - i\langle u, \Omega_Q\rangle\beta\gamma^{-1} - iu\langle\Omega_Q, \beta\gamma^{-1}\rangle + (\frac{b}{a}u - i\langle u\gamma, \beta\rangle\beta\gamma^{-1})\Theta_Q$$

(21) $$\Phi_Q = (d\beta)\gamma^{-1} + \beta\gamma^{-1}d\log\overline{a} + \frac{b}{a}\Omega_Q - i\langle\beta, \Omega_Q\gamma\rangle\beta\gamma^{-1} - i\langle\beta, \beta\rangle\beta\gamma^{-1}\Theta_Q$$

(22) $$\Psi_Q = -2\frac{db}{a} - 2(\frac{b}{a} + i\langle\beta, \beta\rangle)\,d\log\overline{a} - 2i\langle d\beta, \beta\rangle - 2\frac{b}{a}i\langle\Omega_Q, \beta\gamma^{-1}\rangle + 2\frac{b}{a}i\langle\beta\gamma^{-1}, \Omega_Q\rangle$$
$$+ 2((\frac{b}{a})^2 + i\frac{b}{a}\langle\beta, \beta\rangle - \langle\beta, \beta\rangle^2)\Theta_Q$$

We can easily find the structure equation by calculating at the identity element and using the left-invariance. We thus find

$$u\,d\tau_Q = u\tau_Q \wedge \tau_Q - i\langle u, \Phi_Q\rangle \wedge \Omega_Q + iu\langle\Omega_Q, \Phi_Q\rangle + i\langle u, \Omega_Q\rangle \wedge \Phi_Q + \frac{1}{2}\Theta_Q \wedge \Psi_Q,$$

$$d\Phi_Q = \Phi_Q \wedge \tau_Q + \Pi_Q \wedge \Phi_Q + \frac{1}{2}\Omega_Q \wedge \Psi_Q$$

$$d\Psi_Q = \Pi_Q \wedge \Psi_Q + 2i\langle\Phi_Q, \Phi_Q\rangle$$

The above shows that these Pfaffian forms satisfy the conditions (1) to (4) in §5 where all the curvature forms are zero. In view of (10) and the uniqueness of Cartan connection forms proven in §5. We have proven the following

(23) **PROPOSITION.** *The curvatures of* Q *are zero and*

$$\iota^*\Theta = \Theta_Q, \qquad \iota^*\Omega = \Omega_Q, \qquad \iota^*\Pi = \Pi_Q$$

$$\iota^*\tau = \tau_Q, \qquad \iota^*\Phi = \Phi_Q, \qquad \iota^*\Psi = \Psi_Q$$

where $\Theta_Q, \ldots, \Psi_Q$ are the left-invariant Pfaffian forms on B determined by

$$(\Theta_Q)_I = (dx)_I, \qquad (\Omega_Q)_I = (dz)_I, \qquad (\Pi)_I = -(d|a|^2)_I$$

$$(\tau_Q)_I = (d\gamma - da)_I, \qquad (\Phi_Q)_I = (d\beta)_I, \qquad (\Psi_Q) = -2\mathfrak{R}(db)_I$$

The above suggests that we can put more structures on the bundle E of the Cartan connection for any general CR structure M (with the condition on its Levi-form). Namely, we shift our position and denote by (p, t, v, s) a general element in $\mathbb{R}\times GL(n-1, \mathbb{C})\times\mathbb{C}^{n-1}\times\mathbb{R}$ satisfying the condition (8). Denote by P the manifold of also such (p, t, v, s). We regard (a, b, β, γ) representing $T_{(a,b,\beta,\gamma)}$ (mod $\mathbb{Z}_{n+1}$) $\in H = H^*/\mathbb{Z}_{n+1}$. Then the proof of (11) shows that the formula (7) defines a diffeomorphism

$$(24) \qquad \iota_H : H \longrightarrow P$$

Therefore we can put a Lie group structure on P so that ι_H is an isomorphism of Lie groups. By (18) §6 we find that the multiplication in P is given by

$$(p, t, v, s)\cdot(p_1, t_1, v_1, s_1) = (\widetilde{p}, \widetilde{t}, \widetilde{v}, \widetilde{s}),$$

$$(25) \qquad \begin{aligned} \widetilde{p} &= p\rho_1 \\ \widetilde{t} &= tt_1 \\ \widetilde{v} &= \frac{1}{p_1} vt_1 + v_1 \\ \widetilde{s} &= \frac{s}{p_1} + s_1 - 2\Re i\langle v_1 t_1^{-1}, v\rangle \end{aligned}$$

We let H act on P from the right: For $\mathbf{p}\in P$ and $\mathbf{h}\in H$

$$(26) \qquad \mathbf{p}\cdot\mathbf{h} = \mathbf{p}\cdot(\iota_H \mathbf{h})$$

For a CR structure M denote by **c** a choice of $\theta_0, \omega_0, \pi_0, \tau_0, \varphi_0$ as in (0.1) and (0.2) in §2. Then, as was done in §2, **c** induces a fiber chart of the bundle of E, the value of which is in P. For (p, t, v, s) in P we denote by $(p, t, v, s)_{\mathbb{C}}$ the element in E (over a fixed point in M). By (2)′ and (5) in §2, we have

$$(27) \qquad P \ni (p, t, v, s) \longmapsto (p, t, v, s)_{\mathbb{C}} = (\theta, \omega, \pi) \in E$$

where

$$(28) \qquad \begin{aligned} \theta &= p\theta_0 \\ \omega &= \omega_0 t + v\theta \\ \pi &= -d\log\rho + \pi_0 - i\langle v, \omega\rangle + i\langle\omega, v\rangle + s\theta. \end{aligned}$$

We note that, in the above, π is an element in T^*E_2 (cf. (1)′ §2) and p in d log p is the coordinate function p in $(p, t, v, s)_{\mathbb{C}}$. We now let H act from the right on E by means of the above chart in (26). Namely,

$$(29) \qquad (p, t, v, s)_{\mathbb{C}}\cdot\mathbf{h} = ((p, t, v, s)\cdot\mathbf{h})_{\mathbb{C}}$$

Our point is that this action is independent of the choice of $\mathbb{C}$. This will be confirmed when we show that a change of $\mathbb{C}$ induces a change of chart given by the left-multiplication by an element in H. Let $\widetilde{\mathbb{C}}$ be a new choice given by $\widetilde{\theta}_0, \widetilde{\omega}_0, \widetilde{\pi}_0, \widetilde{\tau}_0, \widetilde{\varphi}_0$. Then

(30)
$$\widetilde{\theta}_0 = \rho_* \theta_0$$
$$\widetilde{\omega}_0 = \omega_0 t_* + v_* \widetilde{\theta}_0$$

As was done in §2 it then follows that

$$(t_*)^* t_* = p_*$$

$$d\widetilde{\theta}_0 = i\langle \widetilde{\omega}_0, \widetilde{\omega}_0\rangle + \widetilde{\theta}_0 \wedge (-d\log p_* + \pi_0 - i\langle v_*, \widetilde{\omega}_0\rangle + i\langle \widetilde{\omega}_0, v_*\rangle)$$

Therefore

(31)
$$\widetilde{\pi}_0 = -d\log p_* + \pi_0 - i\langle v_*, \frac{1}{p_*}\widetilde{\omega}_0\rangle + i\langle \frac{1}{p_*}\widetilde{\omega}_0, v_*\rangle + s_* \widetilde{\theta}_0$$

(28) and (30) give the change of chart formula. After calculation we first find that for $(p, t, v, s)_{\mathbb{C}} = (\widetilde{p}, \widetilde{t}, \widetilde{v}, \widetilde{s})_{\widetilde{\mathbb{C}}}$

$$p = p_* \widetilde{p}$$
$$t = t_* \widetilde{t}$$
$$v = \frac{1}{\widetilde{p}} v_* \widetilde{t} + \widetilde{v}$$
$$s = \frac{1}{\widetilde{p}} s_* + \widetilde{s} - 2\mathrm{Re}\, i\langle \widetilde{v}\widetilde{t}^{-1}, v_*\rangle$$

Therefore we find by (25)

(32)
$$(p, t, v, s) = (p_*, t_*, v_*, s_*)(\widetilde{p}, \widetilde{t}, \widetilde{v}, \widetilde{s})$$

The above shows that any change of chart as above is given by a left-multiplication of an element in H. Therefore the bundle of Cartan connection E over M is considered as a principal bundle with structure group H. For $T_{(a,b,\beta,\gamma)} \in H$ and $(p, t, v, s)_{\mathbb{C}} \in E$

$$(p, t, v, s)_{\mathbb{C}} T_{(a,b,\beta,\gamma)} = (p', t', v', s')_{\mathbb{C}}$$

where

$$p' = p|a|^2$$
$$t' = ta\gamma^{-1}$$

$$v' = (\bar{a})^{-1} v\gamma^{-1} - \beta\gamma^{-1}$$

$$s' = \frac{s}{|a|^2} + 2\Re\frac{b}{a} + 2\Re i\langle \frac{1}{a}\beta, v\rangle$$

REFERENCE

[1] S. S. Chern and J. K. Moser, *Real Hypersurfaces in Complex Manifolds,* **Acta Math.** 133 (1974), 219-271.

Department of Mathematics
Columbia University
New York, N.Y. 10027
U.S.A.

H ROSSI

Applications of complex analysis to group representations

1. Introduction

The central problem of representation theory is to understand the structure of a group in terms of its realizations as groups of operators on a vector space. In the case of a Lie group G one restricts attention to *unitary representations:* a continuous homomorphism ρ of G into the group U(H) of unitary transformations of a Hilbert space. ρ is called *irreducible* (ρ is an IUR) if H admits no proper closed invariant subspace. The general theory provides a decomposition (by a direct sum or direct integral) of a unitary representation in terms of IUR's; thus this problem amounts to finding all IUR's. There are two aspects to this problem: i) an extensive aspect —to give an effective parametrization of the set of all IUR's, and ii) an intensive aspect —to provide algorithms which effectively construct these IUR's. Complex analysis plays an important role in this second aspect.

There are two important ways of constructing IUR's. First, by appropriately extending an IUR on a subgroup (induction), and secondly, by cutting down a given representation by imposing conditions (such as holomorphicity). The procedure which is of interest here is a combination of the two, called *holomorphic induction.* First, we shall do some examples in detail which indicate the naturality of this procedure and illustrate the central ideas and problems. Then we shall describe the procedure of holomorphic induction in the abstract, proving that it does produce irreducible representations when they are unitarizable (and non-zero).

2. Examples

2.1. $G = \mathrm{Aff}^+(1)$, the group of orientation-preserving transformations of R^1:

$$G = \{g = (t, x_0);\ t > 0,\ x_0 \in R\}$$

acting on R^1 by

$$g(x) = e^t(x + x_0).$$

We can have G act as a group of linear transformations on a vector space by considering the contravariant action on functions:

$$(g\cdot f)(x) = f(g^{-1}(x)) = f(e^{-t}x - x_0). \tag{2.2}$$

In this realization we do not see an easy way to restrict this to a unitary action on a Hilbert space, but if we separate the action of the variables, it becomes easier. We replace x by $z = x + iy$ and consider the action of G on the upper half plane H^+ by

$$g(z) = e^t(z+x_0).$$

Now, the measure $y^{-2}dxdy$ is invariant under dilations and translations in the x-direction, so G acts by unitary transformations on $L^2(y^{-2}dxdy)$. Furthermore, since the action of G is holomorphic, G acts unitarily by (2.2) on the Hilbert space of holomorphic functions on $L^2(y^{-2}dxdy)$.

Unfortunately this space has only the zero vector: since $y^{-2}dxdy$ is not integrable at $y=0$, we must have

$$\lim_{y\to 0} f(x+iy) = 0$$

in $L^2(dx)$ for such an f. But a holomorphic function with zero boundary values is zero. The principle to follow here is this: whenever you come across a natural and promising construction which nevertheless fails, do not give up, it probably belongs to a family some of whose members will work. Thus, we introduce the family of Hilbert spaces

(2.3) $$H^n = \{f \in (H^+): \|f\|_n^2 = \int |f|^2 y^{n-2} dxdy < \infty\}.$$

Let us look at the contravariant action of G on this new norm:

$$\|g\cdot f\|_n^2 = \int |f(e^{-t}z-x_0)|^2 y^{n-2} dxdy.$$

Letting $w = u+iv = e^{-t}z-x_0$, this becomes

$$\|g\cdot f\|_n^2 = \int |f(w)|^2 (e^t v)^{n-2} e^t du\, e^t dv$$

$$= e^{nt} \|f\|_n^2.$$

Since the contravariant action fails to be an isometry by a scalar factor, this is easy to correct: define the representation ρ_n of G in $U(H^n)$ by

(2.4) $$(\rho_n(g)\cdot f)(z) = e^{-nt/2} f(e^{-t}z-x_0)$$

for $g=(t, x_0)$.

2.5. **Proposition.** $H^n = \{0\}$ if $n>1$.

Proof. Let ϕ be a function on R^+, and define

(2.6) $$f(z) = \int_0^\infty e^{iz\xi}\phi(\xi)d\xi = (e^{-y\xi}\phi)\hat{}(x),$$

where "^" means "Fourier Transform". Then

$$\|f\|_n^2 = \int_0^\infty [\int_0^\infty |f(x+iy)|^2\, dx]\, y^{n-2}\, dy$$

$$= \int_0^\infty [\int_0^\infty e^{-2y\xi} |\phi(\xi)|^2\, d\xi]\, y^{n-2}\, dy$$

by the Plancherel formula. Now, using the Fubini theorem, we get

$$\|f\|_n^2 = \int_0^\infty |\phi(\xi)|^2 [\int_0^\infty e^{-2y\xi}\, y^{n-2}\, dy]\, d\xi$$

$$= \frac{\Gamma(n-1)}{2^{n-1}} \int_0^\infty |\phi(\xi)|^2\, \xi^{-(-n-1)}\, d\xi.$$

The last equation is valid only for $n>1$. In this case, for a continuous ϕ of compact support, not identically zero, f is a nonzero function in H^n.

In fact the proposition holds if and only if $n>1$ (our argument that $H^0=\{0\}$ works for all $n \leqslant 1$). The integral (2.6) in fact defines an isometry (but for the scale factor $2^{1-n}\Gamma(n-1)$) between H^n and $L^2(R^+;\, \xi^{-(n-1)} d\xi)$. Using this isometry and abelian harmonic analysis it can be shown that the representations ρ_n are irreducible (and in fact all equivalent).

(2.7) $$G = SL(2,R) = \{g = \begin{pmatrix} a & b \\ c & d \end{pmatrix};\ ad-bc=1\}.$$

G is described as a group of linear transformations of R^2 and the contravariant action on $L^2(R^2)$ is unitary, but we soon find out this representation is difficult to analize. The picture becomes manageable if, as above, we turn instead to the action on $\mathbb{C}^2$

$$g\begin{pmatrix} u \\ v \end{pmatrix} = \begin{pmatrix} au+bv \\ cu+dv \end{pmatrix},\ \begin{pmatrix} u \\ v \end{pmatrix} \in \mathbb{C}^2$$

We can identify G with the orbit Σ of $\begin{pmatrix} -1 \\ 1 \end{pmatrix}$, since the action of G is faithful. If $d\mu$ is Haar measure on G, the contravariant action of G is unitary in $L^2(\Sigma, d\mu)$ and the space of holomorphic functions is invariant. However, this space is not a Hilbert space, so we take its L^2-closure:

(2.8) $$H(G) = \{\text{limits in } L^2(d\mu) \text{ of functions holomorphic in } \mathbb{C}^2\}.$$

These are the square-integrable CR-functions, where for the purpose of these lectures, we use this definition.

2.9. **Definition.** *Let* Σ *be a real submanifold of a complex manifold* M. *A* CR-function *on* Σ *is a function* f *which (locally) satisfies this condition:* if ϕ is a distribution on Σ such that $\langle h, \phi\rangle = 0$ for all h holomorphic (locally) on M, then $\langle f, \phi\rangle = 0$ also.

We can analize $H(G)$ most easily by making the variable change

$$z = \frac{u}{v} \qquad w = v^{-1}.$$

We calculate the action of G in these coordinates. Let $(u', v') = g(u, v)$, so that

$$u' = au+bv = [a(\frac{u}{v}) + b]v$$

$$v' = cu+dv = [c(\frac{u}{v}) + d]v.$$

In the new coordinates we have

$$g(z, w) = (\frac{az+b}{cz+d}, (cz+d)^{-1} w). \tag{2.10}$$

The surface Σ is given by the equation $|w|^2 = \mathrm{Im}\, z$. Thus the projection $(z, w) \longrightarrow z$ (which is the projection $\mathbb{C}^2 \longrightarrow P^1$) exhibits Σ as a circle bundle over the upper half plane H^+ (lying in the dual of the hyperplane bundle over P^1). This is a Hartogs domain in (z, w)-space, so every holomorphic function admits a Laurent expansion:

$$f(z, w) = \sum_{-\infty}^{\infty} f_n(z) w^n; \; f_n \in \theta(H^+).$$

This is proven as follows: for each $z \in H^+$, (2.10) is the Fourier expansion of F on the circle $|w|^2 = \mathrm{Im}\, z$. If f is holomorphic in a neighborhood of Σ, the coefficients remain in $\theta(H^+)$. Now the action of G is linear in w, so that decomposition (2.10) is preserved by G. More precisely,

$$H(G) = \oplus H^n \quad \text{where} \quad H^n = f(z)w^n \in H(G)\,. \tag{2.11}$$

Writing $z = x+iy$, $w = \sqrt{y}\, e^{i\theta}$, Haar measure becomes $y^{-2} dx dy d\theta$, and we find that H^n is the same H^n as defined in (2.4):

$$\iiint |f(x+iy)(\sqrt{y}\, e^{i\theta})^n|^2\, y^{-2} dx dy d\theta = 2\pi \iint |f(x+iy)|^2\, y^{n-2} dx dy$$

Thus we have proven:

2.12. **Proposition.** G *acts unitarily on the space* H(G) *of* L^2(d Haar) CR-*functions defined in* (2,8). *We have the decomposition into irreducible components*

$$H(G) = \bigoplus_{n \geq 2} H^n$$

where H^n is defined by (2.4).

The action is irreducible, since the $\mathrm{Aff}^1(1)$-action is contained in that of G: $\mathrm{Aff}^+(1)$ is (isomorphic to) the subgroup of SL(2, R) of matrices of the form

$$\begin{bmatrix} e^{t/2} & e^{t/2} x_0 \\ 0 & e^{-t/2} \end{bmatrix} = \begin{bmatrix} e^{t/2} & 0 \\ 0 & e^{-t/2} \end{bmatrix} \begin{bmatrix} 1 & x_0 \\ 0 & 1 \end{bmatrix}$$

Since the $\mathrm{Aff}^+(1)$-action is irreducible so is that of $SL(2, R)$.

2.13. $G = SU(2)$, the group of 2×2 unitary matrices of determinant one. G acts as a group of holomorphic transformations on $\mathbb{C}^2$, and the unit sphere Σ is an orbit under G. In fact,

$$SU(2) = \begin{pmatrix} a & -\bar{b} \\ b & \bar{a} \end{pmatrix}; \ |a|^2 + |b|^2 = 1$$

and the identification of Σ with G:

$$\begin{pmatrix} a \\ b \end{pmatrix} \longrightarrow \begin{pmatrix} a & -\bar{b} \\ b & \bar{a} \end{pmatrix}$$

commutes the action of $SU(2)$ on 2-vectors with the left action on itself. Thus, if $d\theta$ is left Haar measure (ordinary Euclidian surface area on the sphere) we get a unitary representation of $SU(2)$ on

$$H(\Sigma) = \{L^2(d\theta)\text{-limits of holomorphic functions}\}. \tag{2.14}$$

All such functions are boundary values of functions holomorphic on the ball, so admit Taylor expansions

$$f(z) = \sum_{n\geq 0} f_n(z)$$

where $f_n(z)$ is a homogeneous polynomial of degree n. The expansion (2.15) is invariant under the $SU(2)$-action.

2.15. **Proposition.** G *acts unitarily on the subspace* $H(\Sigma)$, (2.14), *of* $L^2(G)$ *of boundary values of holomorphic functions under the identification* $G \leftrightarrow \Sigma$. *We have the decomposition into irreducible components* $H(\Sigma) = \bigoplus_{n\geq 0} H_n$, *where* H^n *is the space of homogeneous polynomials of degree* n.

2.16. It is a classical fact that these representations are irreducible; this will follow from a theorem we will prove below.

2.17. $G = SU(2, 1)$.

Let w_1, w_2, w_3 be coordinates in $\mathbb{C}^3$. G is the group of linear transformations in $\mathbb{C}^3$ leaving the form

$$\phi(w) = |w_1|^2 + |w_2|^2 - |w_3|^2$$

invariant. The orbit of the point (0, 0, 1) is the surface

$$\Sigma: \ \phi(w) = -1,$$

and it carries a G-invariant measure $d\mu$. Again, we take as our Hilbert space the $H(\Sigma)$ of $L^2(d\mu)$-

limits of holomorphic functions: the square-integrable CR-functions on Σ. (Since Σ has Levi signature (1, 1), these functions are in fact holomorphic in a neighborhood of Σ).

To further analize $H(\Sigma)$ we make the change of variables

$$z_1 = w_1 w_3^{-1}, \qquad z_2 = w_2 w_3^{-1}, \qquad z_3 = w_3^{-1}.$$

The map $\pi = (z_1, z_2): \Sigma \longrightarrow \mathbb{C}^2$ exhibits Σ as a circle bundle over the unit ball B in $\mathbb{C}^2$, since Σ is the surface

$$|z_3|^2 = 1 - (|z_1|^2 + |z_2|^2)$$

Again, we have a Taylor expansion (in z_3) for functions in $H(\Sigma)$, and degrees are preserved. Thus if

$$H_n = \{f \in \mathcal{O}(B);\ f(z_1, z_2)\, z_3^n \in H(\Sigma)\},$$

we have a decomposition $H(\Sigma) = \bigoplus_{n \geq 0} H_n$, and the H_n are irreducible. We shall not pursue the details further.

If we apply the same procedure to any surface $\phi(w) = c$, so long as $c < 0$, we get the same set of representations; these are all in the holomorphic discrete series. For $c = 0$ or $c > 0$, we get, with a more difficult analysis, a different set of representations.

Note that the map $\pi: \Sigma \longrightarrow \mathbb{C}^2$ is just the projection map of $\mathbb{C}^3 \longrightarrow P^2$ given in inhomogeneous coordinates of P^2.

2.18. $G = SU(2, 2)$.

Let $w_1, \ldots, w_4$ be coordinates in $\mathbb{C}^4$. G is the group of linear transformations of determinant one leaving the form

$$\phi(w) = |w_1|^2 + |w_2|^2 - |w_3|^2 - |w_4|^2$$

invariant. This example displays a much more complex geometric involvement than the preceding. As above, the orbit of the point (0, 0, 0, 1) is the surface

$$\Sigma\colon \phi(w) = -1,$$

which also carries a G-invariant measure, leading to a unitary representation on the space $H(\Sigma)$ of square-integrable CR-functions.

To analize $H(\Sigma)$, we introduce the natural projection $\pi: \mathbb{C}^4 \longrightarrow P^3$; again Σ is a circle bundle over a domain P^3_- in P^3, and we anticipate a decomposition into spaces of holomorphic functions on P^3_-. But here the analogy with preceding cases breaks down: P^3_- is not at all like the ball, and we show as a consequence that $H(\Sigma) = 0$. In the coordinate neighborhood $w_4 \neq 0$, P^3_- has the equation

$$|z_1|^2 + |z_2|^2 - |z_3|^2 < 1$$

so has less signature (1, 1). Thus P^3_- is 1-concave. If H had any nonzero functions, the coefficients in the Laurent expansion on the fibers of π would be sections of some line bundle over P^3_- (in fact, powers of the hyperplane section bundle). Now the space of sections of such a bundle is finite-dimensional and would produce a finite-dimensional unitary representation of SU(2, 2). But there are no such representations.

We cannot cure this situation by modifying the choices made on the way; we will never get representations on spaces of holomorphic functions. However, since a 1-concave domain admits infinite-dimensional first cohomology spaces, we may be able to find unitary representations of SU(2, 2) in $H^1(P^3_-, L)$ for good choices of line bundles L. This is in fact the case, and was first discovered in the study of the role SU(2, 2) plays in quantum physics. We shall describe the construction of Roger Penrose (see [15]) which brings this out.

We can view P^3_- as the set of lines in $\mathbb{C}^4$ on which the form ϕ is negative definite. Let M_- be the set of planes in $\mathbb{C}^4$ on which ϕ is negative definite. M_- is an open subset of the manifold Gr(4, 2) of planes in $\mathbb{C}^4 = \mathbb{C}^2 \oplus \mathbb{C}^2$; we can parametrize an open subset N of Gr(4, 2) by $\mathrm{Hom}(\mathbb{C}^2, \mathbb{C}^2) = \mathbb{C}^{2\times 2}$; for $\tau \in \mathbb{C}^{2\times 2}$, let P_τ be the plane

$$P_\tau = \left\{\begin{pmatrix} \tau v \\ v \end{pmatrix};\ v \in \mathbb{C}^2\right\}$$

(P_τ is the graph of τ). For $g \in GL(4, \mathbb{C})$, write g in block form relative to this direct sum:

$$g = \left\{\begin{pmatrix} A & B \\ C & D \end{pmatrix}:\ A, B, C, D \in \mathbb{C}^{2\times 2}\right\}$$

Then

$$g \cdot P_\tau = \left\{\begin{pmatrix} (A\tau + B)v \\ (C\tau + D)v \end{pmatrix}\right\} = \left\{\begin{pmatrix} (A\tau + B)(C\tau + D)^{-1} w \\ w \end{pmatrix}\right\}$$

where $w = (C\tau + D)v$. Thus

$$g \cdot P_\tau = P_{\tau'}$$

where

(2.19) $$\tau' = (A\tau + B)(C\tau + D)^{-1},$$

if τ' is in the same coordinate neighborhood N (otherwise $C\tau + D$ need not be invertible). A calculation verifies that $M_- \subset N$, and the condition that $P_\tau \in M_-$ comes down to the realization

$$M_- = \{\tau \in \mathbb{C}^{2\times 2};\ \frac{1}{i}(\tau - {}^t\bar{\tau}) >> 0\},$$

with SU(2, 2) acting transitively by the fractional linear action (2.19). Now M_- is a Stein domain, and what we probably wanted was a map of $\mathbb{C}^4$ to Gr(4, 2). None such exists, so we replace $\mathbb{C}^4$ by

$$S = \{(v_1, v_2) \in \mathbb{C}^4 \times \mathbb{C}^4;\ v_1 \text{ and } v_2 \text{ are independent}\},$$

and now the fibration π: S $\longrightarrow$ Gr(4, 2) is defined:

$$\pi(v_1, v_2) = \text{span}(v_1, v_2),$$

Let $S = \{(v_1, v_2);\ \phi(v_1) < 0,\ \phi(v_2) < 0\}$. Then $\pi: S_- \longrightarrow M_-$. The generic SU(2, 2)-orbit Σ' is a G-invariant measure. The space $H(\Sigma')$ of square-integrable CR-functions is the Hilbert space we seek. If $f \in H(\Sigma')$, f has a Taylor expansion

$$f(v_1, v_2) = \Sigma f_n(v_1, v_2)$$

where f_n is homogeneous of degree n on the fiber. Thus $H(\Sigma') = \oplus H_n$, where the H_n are Hilbert spaces of holomorphic functions on M_-, and G-invariant. As we shall see below, these representations are irreducible.

Now, we see how Penrose relates these representations with cohomology on P^3_-. Let

$$p: S_- \longrightarrow P^3_-,\ p(v_1, v_2) = \text{span}(v_1).$$

For $\tau \in M_-$, $p\pi^{-1}(\tau) = V_\tau$ is a 1-dimensional complex submanifold of P^3 (biholomorphic to P^1 in fact). If $L \longrightarrow P^3_-$ is a holomorphic vector bundle, let

$$H_{L,\tau} = H^1(V_\tau, L).$$

$UH_{L,\tau} = H_L$ can be given the structure of a holomorphic vector on M_-. These vector bundles are related as G-spaces to the Hilbert spaces of holomorphic functions described above. Finally, the association $\phi \longrightarrow \sigma$ of a cohomology class in $H^1(P^3, L)$ with a section of H_L is defined by

$$\sigma(\tau) = \phi | V_\tau.$$

This association is the "Penrose correspondence" relating cohomology spaces on P^3 with members of the holomorphic series on M_- as G-spaces.

These examples are illustrations of the technique of "holomorphic induction", which plays a central role in the representation theory of many families of groups. In the next section we shall give the abstract formulation of this process and prove a theorem of Kunze: *a representation holomorphically induced from an irreducible representation is itself irreducible.*

I do not want to leave the impression that this is all "semisimple theory": theta function theory arises from such structures on step two nilpotent groups, and the Auslander-Kostant theory of IUR's for type I solvable group depends crucially on this construction.

III. Homogeneous Complex Structures on Groups

Let G be a Lie group: a topological group which has the structure of a real analytic manifold so that the group operations are real analytic. We thus get two homomorphisms of G into the group of diffeomorphisms of G, called the *left-action* L_g, and *right-action* R_g, respectively:

$$L_g(h) = g^{-1}h, \qquad R_g(h) = hg.$$

Note that the left and right actions commute. By "Haar measure" on G we will mean the (unique) left-invariant measure. Then the left action induces a unitary representation on L^2(d Haar):

$$(g \cdot f)(h) = f(g^{-1}h). \tag{3.1}$$

This representation is called the *(left) regular representation* of G.

We let G act on vector fields by the differential of the left-action. The *Lie algebra* of G, denoted g, is the algebra under the brackets of G-invariant vector fields. As a vector space, g is isomorphic to $T_e(G)$ by

$$g = \{dL_g(v);\ v \in T_e(G)\}.$$

By $g^{\mathbb{C}}$ we mean the complexification of g: the space of G-invariant vector fields with complex coefficients. $g^{\mathbb{C}}$ is the Lie algebra of a complex Lie group $G^{\mathbb{C}}$ (this means that $G^{\mathbb{C}}$ is a complex analytic manifold, and the group operations are holomorphic.

Let h be a subalgebra of $g^{\mathbb{C}}$ such that $h \cap \overline{h} = \{0\}$. For each $x \in G$, the set of values of h at x is a subspace H_x of $T_x(G)^{\mathbb{C}}$, and $H = \cup H_x$ is a subbundle of $T(G)^C$ with these two properties:

(i) $H \cap \overline{H} = \{0\}$

(ii) The space of sections of H is closed under Lie bracket.

Thus, such a subalgebra h gives rise to the abstract CR-manifold (G, H) and, since everything is real-analytic, the structure arises (locally) from an embedding of G as a real hypersurface in a complex manifold. The differential equations

$$\overline{X} \cdot f = 0, \quad X \in h \tag{3.2}$$

are the *induced Cauchy-Riemann equations* and we denote the space smooth of CR-functions by

$$O_h(G) = \{f \in C^\infty(G);\ \overline{X} \cdot f = 0,\ X \in h\}. \tag{3.3}$$

In the context of the above discussion, a CR function was a solution of the CR-equations in the sense of distributions (or see definition 3.6 below).

If $h \oplus \overline{h} = g^{\mathbb{C}}$, then the CR-structure is in fact a structure of complex manifold so that the left-action is holomorphic. Notice that this does not make G a complex Lie group: in a complex Lie group both the left and right actions are holomorphic.

3.4. **Proposition.** *Given a Lie subalgebra* h *of* $g^{\mathbb{C}}$ *such that* $h \cap \overline{h} = \{0\}$, *there is a complex manifold* M *and a faithful action of* G *on* M *so that the CR-structure* (G, H) *is that induced by the embedding of* G *as an orbit in* M.

Proof. If h is the Lie algebra of a closed subgroup of $G^{\mathbb{C}}$, then we take $M = G^{\mathbb{C}}/H$. Even if H is not closed, this works in a neighborhood of the G-orbit of e.

3.5. **Example.** su(2), the Lie algebra of SU(2).

su(2) is generated by vectors A, B, C with [A, B] = C cyclically. Let h be the span of B-iC. Then $G^{\mathbb{C}}/H$ is C^{2*} and SU(2) appears as the unit sphere. In the coordinates (t, z), where

$$w = (1-|z|^2)\, e^{it},$$

the CR-operator is the Hans Lewy operator

$$\frac{\partial}{\partial z} + iz\frac{\partial}{\partial t}.$$

$O_h(G)$ is the space of boundary value of holomorphic functions.

3.6. $H(G, h) = \{f \in L^2(d\mu);\ \overline{X}\cdot f = 0 \text{ weakly, } X \in h\}$

In each of the above examples, this was the first Hilbert space encountered. As we saw, this is usually not irreducible, so we seek invariant subspaces. Here is where induction comes in.

Suppose K is a subgroup of G and the distinguished algebra h is K-invariant (here of course we mean the right-action, since every vector in h is already left-invariant). Precisely, for $k \in K$ and $X \in h$, $dR_k(X)$ is another left-invariant vector field. We assume that if $X \in g$, $dR_k(X) \in h$ also, for all $k \in K$. Then, if χ is any character of K $(\mathrm{Hom}(K, S_1))$,

$$H(G, h, \chi) = \{f \in H(G, h);\ f(xk) = \chi(k)^{-1} f(x)\} \tag{3.7}$$

is a G-invariant subspace of H(G; h).

3.8. **Example.** SU(2). h is invariant under the S_1-action

$$(z, w) \longrightarrow (e^{-i\theta} z, e^{i\theta} w):$$

$$\frac{\partial}{\partial z} + iz\frac{\partial}{\partial t} \longrightarrow e^{i\theta}\left(\frac{\partial}{\partial z} + iz\frac{\partial}{\partial t}\right).$$

If $\chi_n(e^{i\theta}) = e^{in\theta}$, then $H(G, h, \chi_n)$ is the space of homogeneous polynomials on $\mathbb{C}^2$ of degree n. This representation is irreducible.

More generally, if V is any Hilbert space, and ρ a unitary representation of K on V, we can construct $H(G; h, \rho)$ as the set of V-valued square-integrable functions on G which are weak solutions of the CR-equations defined by h and which transform on the right according to ρ:

$$f(xk) = \rho(k)^{-1} f(x).$$

The theorem we shall prove (due to R. Kunze [6]) states: if $k, h, \overline{h}$ span $g^{\mathbb{C}}$, then this representation is irreducible if nonempty, if ρ is irreducible. Before we give the proof we must discuss the existence of holomorphic structures on homogeneous bundles in Lie algebra terms.

3.10. **Definition.** *Let* G *be a Lie group,* K *a closed subgroup of* G *and* X *a differentiable manifold. Suppose* π *is a homomorphism of* K *into the group of diffeomorphisms of* X. *The* G*-homogeneous bundle associated to* ρ *and with fiber* X *is defined as*

$$(G \times X)/E,$$

where E *is the equivalence relation*

$$(gk, x)\, E\, (g, \rho(k)^{-1} x), \quad x \in X, \quad g \in G, \quad k \in K.$$

We shall not verify the following.

3.11. **Proposition.** *The natural projection* $G \times X \longrightarrow G$ *induces a projection* $\pi: G \times_\rho X \longrightarrow G/K$ *exhibiting* $G \times_\rho X$ *as a fiber bundle over* G/K *with fiber* X *and group* $\rho(G)$. *The left action of* G *commutes with* π *so realizes* G *as a group of bundle maps of* $G \times_\rho X$.

Conversely, if $B \longrightarrow G/K$ is a bundle with group H and fiber X on which G acts as a group of bundle maps, then $B \approx G \times_\rho X$ as bundles over G/K and G-homogeneous spaces. Here $\rho: K \longrightarrow H$ is the homomorphism

$$\rho(k)(x) = \phi(k \cdot \phi^{-1}(x))$$

where ϕ is some isomorphism of X with the fiber over eK.

In particular, if X is a vector space V, this proposition exhibits an equivalence between representations $\rho: K \longrightarrow GL(V_x)$ and G-homogeneous X-bundles over G/K. Finally, we can identify sections of the bundle $G \times_\rho V$ with the space of functions on G satisfying (3.9), by pulling back over the equivalence relation.

Suppose now that G/K carries a G-invariant complex structure; i.e., G/K is a complex manifold and the left action of G is given by holomorphic transformations. Such a structure is given by a tensor field J on T(G/K), satisfying $J^2 = -I$. This leads to a decomposition (J extended to $T(G/K)^{\mathbb{C}}$ has two eigenspaces):

(3.12) $$T(G/K)^{\mathbb{C}} = H \oplus \overline{H}$$

where H is the eigenspace of eigenvalue i: the bundle of holomorphic ((1, 0))-vectors. Since the left action of G is holomorphic, this splitting is left-invariant.

The Lie algebra g is pointwise-fixed for the left action of G (by definition) and the right action takes left-invariant fields to left-invariant fields. This representation of G on g is denoted

$$\mathrm{ad}\colon G \longrightarrow GL(g).$$

Since ad/K takes k to k, we have an induced action $\mathrm{ad}\colon K \longrightarrow GL(g/k)$. It is now easy to verify that the correspondence of proposition (3.11) gives us

$$T(G/K) \cong G\times_{\mathrm{ad}} g/k$$

as G-invariant vector bundles. Since the splitting (3.12) is G-invariant, it comes from a splitting

$$(g/k)^{\mathbb{C}} = h \oplus \overline{h} \tag{3.13}$$

where h is ad-K-invariant, since we must have

$$H = G\times_{\mathrm{ad}} h.$$

Pulling back to G, we see that the data on G giving rise to a complex structure consists in a subspace g^+ of g

(3.14)

(i) g^+ is ad-K-invariant

(ii) $g^+ \cap \overline{g^+} \subset k^{\mathbb{C}}$

(iii) $g^+ + \overline{g^+} = g^{\mathbb{C}}$

Then $h = g^+/k^{\mathbb{C}}$ (and g^+ need not be the full inverse image of h). There is one ingredient missing: the space of sections of H is closed under Lie brackets. This translates into the requirement that g^+ be a Lie algebra.

3.15. **Proposition.** *Let* K *be a closed subgroup of* G. *A G-homogeneous complex structure on* G/K *is determined by a Lie subalgebra* g^+ *of* $g^{\mathbb{C}}$ *satisfying* (3.14i-iii).

We shall denote $\overline{g^+}$ by g^-.

3.16. **Example: Compact Groups**

Let G be a compact Lie group and T its maximal abelian subgroup. The adjoint representation $\mathrm{ad}\colon T \longrightarrow GL(g)^{\mathbb{C}}$ is diagonalizable, so that we can write

$$g^{\mathbb{C}} = t^{\mathbb{C}} \oplus \sum_{\alpha\in\Delta} g_\alpha$$

where t is the Lie algebra of T and $\Delta \subset t^{\mathbb{C}k}$ and g_α is the eigenspace (*rootspace*) with eigenvalue (*root*) α:

$$\mathrm{ad(H)(x)} = \alpha(\mathrm{H})\,\mathrm{X} \qquad \mathrm{H}\in \mathrm{t},\ \mathrm{X}\in g_\alpha.$$

It is a fact that the α are purely imaginary, and if α is a root, so is $-\alpha\,(=\overline{\alpha})$ and either $\alpha+\beta$ is a root (for two roots α and β) or $[g_\alpha, g_\beta]=0$. We can write $\Delta=\Delta^+\cup-\Delta^+$ where Δ^+ has the property

(3.17) $$\text{if } \alpha,\beta \in \Delta^+,\ \alpha+\beta \text{ a root, then } \alpha+\beta\in\Delta^+$$

Such a Δ^+ is called a positive rootsystem, and for such a choice, the space $g^+ = \sum_{\alpha\in\Delta^+} g_\alpha$ is a Lie algebra satisfying all the above conditions and thus describes a complex structure on G/T.

More generally, the same structure exists for a semi-simple Lie group G which has a maximal abelian subgroup T which is compact. We can write

$$g^{\mathbb{C}} = \mathrm{t}^{\mathbb{C}}\oplus g^+\oplus g^-, \qquad g^+ = \sum_{\alpha\in\Delta^+} g_\alpha$$

where Δ^+ is a choice of positive root system (satisfying (3.17)). If K is the maximal compact subgroup of G containing T, g^+ is ad-K-invariant, and thus G/K has a G-homogeneous structure. We can replace K by any connected subgroup H such that $\mathrm{H}\supset\mathrm{T}$.

We turn to bundles. Let G be a Lie group, K a closed subgroup and g^+ a Lie subalgebra of $g^{\mathbb{C}}$ defining a complex structure on G/K. Let H be a Lie group carrying a left-invariant complex structure:

(3.18) $$h = h^+\oplus\overline{h^+}$$

(the splitting need not be ad-H-invariant; if it is, H is a complex Lie group). Let $\rho\colon \mathrm{K}\longrightarrow\mathrm{H}$ be a homomorphism defining the homogeneous bundle $\mathrm{G}\times_\rho\mathrm{H}\longrightarrow\mathrm{G/K}$. The differential $d\rho\colon \mathrm{k}\longrightarrow h$ is a Lie algebra homomorphism and splits, according to (3.18), into a (1, 0)- and a (0, 1)-part. The next proposition is a verification like that of (2.15) above.

3.19. **Proposition.** *Given the above data, with* K *connected, the bundle* $\mathrm{G}\times_\rho\mathrm{H}\longrightarrow\mathrm{G/K}$ *is a G-homogeneous holomorphic vector bundle if and only if there is a representation* $\sigma\colon \mathrm{k}^{\mathbb{C}}+g^+\longrightarrow h^+$ *which extends* $d\rho^{(1,0)}$. If K is not connected, we must assume

$$\mathrm{ad}\rho(\mathrm{k})\cdot\sigma(\mathrm{X}) = \sigma((\mathrm{ad\,k})(\mathrm{X})) \qquad \mathrm{k}\in\mathrm{K},\ \mathrm{X}\in g^+.$$

This is an equivalence of categories: two holomorphic bundles are equivalent (as G-bundles) if and only if their σ's are equivalent representations.

It is of interest to apply to the injection $\mathrm{K}\longrightarrow\mathrm{K}^{\mathbb{C}}$: the bundle $\mathrm{G}\times_\mathrm{K}\mathrm{K}^{\mathbb{C}}\longrightarrow\mathrm{G/K}$ carries a complex structure if and only if the injection $\mathrm{k}\longrightarrow\mathrm{k}^{\mathbb{C}}$ extends to a homomorphism $\sigma\colon \mathrm{k}^{\mathbb{C}}+g^+\longrightarrow\mathrm{k}^{\mathbb{C}}$. If $\mathrm{m}^+=\ker\sigma$, then $\mathrm{k}^{\mathbb{C}}+g^+=\mathrm{k}^{\mathbb{C}}\oplus\mathrm{m}^+$ and m^+ defines the complex structure on G/K. Thus

3.20. **Proposition.** $G\times_K K^{\mathbb{C}} \longrightarrow G/K$ *is a holomorphic* G*-homogeneous bundle over* G/K *if and only if*

(3.21) $$g^+ = k^{\mathbb{C}} \oplus m^+$$

where m^+ is an ideal in g^+.

In this case, every homomorphism $d\rho^{(1,0)}$ of $k^{\mathbb{C}}$ extends to g^+, so we have

3.22. **Corollary.** *When we have* (3.21), *every bundle over* G/K *whose group is a complex Lie group carries a* G*-invariant complex structure.*

Notice that G can be realized as any orbit in $G\times_K K^{\mathbb{C}}$, giving a direct and explicit proof of Proposition 3.4 in this case.

At the other extreme, let $f \in g^*$. The infinitesimal condition for the equation

(3.23) $$\chi(\exp K) = \exp(i\langle f, K\rangle),$$

$K \in k$ to define a homomorphism of K to C^* is that $f([k, k]) = 0$. In this case $f = d\chi$, and χ defines a line bundle $L_\chi \longrightarrow G/K$. If

(3.24) $$f([g^+, g^+]) = 0,$$

f extends $d_\chi^{1,0}$ as a homomorphism to g^+, so L_χ is a holomorphic line bundle when (3.24) is satisfied. In this case g^+ is said to be a *complex polarization* at f. This concept is crucial in the coadjoint orbit theory of representations ([2], [5]).

3.25. **The Borel-Weil Theorem**

In the case of a semi-simple Lie group, the sum g^+ of the positive root spaces is an ideal in $k^{\mathbb{C}} \oplus g^+$ since it is ad(K)-invariant, so qualifies as the m^+ in (3.21). Thus every complex vector bundle over G/K (or G/T) carries a holomorphic structure.

Let G be a compact group with maximal torus T and let $r = \dim T$. Then $G\times_T T^{\mathbb{C}} \longrightarrow G/T$ is a homogeneous holomorphic vector bundle ($T^{\mathbb{C}} \cong (C^*)^r$ and T lies in $(C^*)^r$ as the distinguished boundary of the polydisc). With the proper choice of Δ^+, this vector bundle is exceptional, so there are many holomorphic functions on it, and we can describe H(G) as the closure in L^2 of Haar measure of the space of functions holomorphic on V. Each such function has a Taylor expansion and since G acts as $(C^*)^r$ on the fibers the Taylor decomposition is preserved. Precisely, for $\chi \in \hat{T}$, $\chi = (m_1, \ldots, m_n)$ we take

$$H(G, \chi) = \{f(z) w_1^{m_1} \ldots w_n^{m_n} \in H(G)\}$$

where z is the base coordinate and w the fiber coordinate. Then

(3.26) $$H(G) = \bigoplus_{\chi \in \hat{T}} H(G, \chi)$$

and the decomposition is G-invariant. If $E_\chi \longrightarrow G/T$ is the line bundle associated to the character χ, $H(G, \chi) \cong H^0(G/T, E_\chi)$ in a natural way.

The Borel-Weil theorem asserts that each IUR of G occurs once and only once in (3.26), and furthermore describes the set of χ which actually appear (those χ, with $H(G, \chi) \neq \{0\}$).

If we apply the same construction to a semi-simple Lie group G with compact maximal abelian subgroup T, we find that G is an orbit in $G \times_K K^{\mathbb{C}} \longrightarrow G/K$ and we can take the fiber to be the vector bundle $V \longrightarrow K/T$. In this case H(G) decomposes according to the K-decomposition of H(K) on the fiber; these are the holomorphically-induced representations. Not all are non-zero, those which are comprise the *discrete holomorphic series* of G.

3.27. The Irreducibility Criterion

3.28. **Definition.** *Let* K *be a subgroup of a Lie group* G *and* $\rho: K \longrightarrow U(V)$ *a unitary representation of* K *on a vector space* V. *The contravariant action of* G *on the space of sections* $C^\infty(G/K, G\times_\rho V)$ *is the* induced representation *of* G. *If* g^+ *is a Lie algebra defining a complex structure on* G/K, *and* $d\rho^{1,0}$ *extends to a homomorphism of* $k^{\mathbb{C}}+g^+$, $G\times_\rho V$ *is a holomorphic vector bundle* $L \longrightarrow G/K$. *The contravariant action of* G *on the space of holomorphic sections is the* holomorphically-induced representation *of* G.

3.29. **Proposition.** *Suppose we are in the above situation, and* H *is a non-empty Hilbert space of sections of* L *on which* G *acts unitarily. Suppose in addition that*

(1) ρ *is irreducible*

(2) *for all* $z \in Z$, *the evaluation map* $E_z: f \longrightarrow f(z)$ *is a continuous map of* H *into* $\pi^{-1}(z)$.

Then the representation of G *on* H *is irreducible.*

Proof. Let R represent the contravariant representation of G on H; we want to show that H has no proper closed subspaces which are R-invariant. For $g \in G$, let $\sigma(g, z): L_{g^{-1}z} \longrightarrow L_z$ denote the fiber mapping given by left translation by g. Then R is given by

$$(R(g)\cdot f)(z) = \sigma(g, z)\, f(g^{-1} z). \tag{3.30}$$

and ρ is given by

$$\rho(k) = \sigma(k, z_0)$$

where z_0 is the point isotropy group K. We introduce the "Kernel function"

$$Q(z, w): L_w \longrightarrow L_z, \qquad Q(z, w) = E_z E_w^*.$$

This is the Kernel function, for if $Q_w \phi = E_w^* \phi$, then $(Q_w\phi)(z) = Q(z, w)\phi$ and $\langle \psi, \phi\rangle_{L_w} = \langle Q_w \psi, Q_w \phi\rangle_H$
Clearly

$$Q(z,w)^* = Q(w,z)$$

and $Q(z,z) > 0$ as an operator on L_z (because G acts transitively on the base and K irreducibly on the fiber, it is easy to see that E_z is surjective for all $z \in G/K$). Now (3.30) is restated as

$$E_z R(g) = \sigma(g,z) E_{g^{-1}z}.$$

Now, for all g, R(g) is an isometry, so $R(g)^{-1} = R(g)^*$. We obtain

(3.31) $$Q(z,w) = E_z R(g)(E_w R(w))^* = \sigma(g,z) Q(g^{-1}z, g^{-1}w) \sigma(g,w)^*$$

for all $g \in G$, and $z, w \in K$. Taking $z = w = z_0$, this gives

$$Q(z_0, z_0) = \rho(k) Q(z_0, z_0) \rho(k^{-1})$$

for $k \in K$. Thus $Q(z_0, z_0)$ commutes with the irreducible representation ρ of K so, by Schur's lemma, $Q(z_0, z_0)$ is a scalar operator.

Now suppose that H′ is a closed non-zero invariant subspace of H. The entire preceding discussion holds for H′ and its Kernel function Q′. Thus $Q(z_0, z_0)$ and $Q'(z_0, z_0)$ are scalar operators, so

$$Q(z_0, z_0) = cQ'(z_0, z_0)$$

for some $c > 0$. By (3.31), for both Q and Q′, and $z = g(z_0)$

$$\begin{aligned} Q(z,z) &= \sigma(g,z) Q(z_0,z_0) \sigma(g,z)^* \\ &= \sigma(g,z) cQ'(z_0,z_0) \sigma(g,z)^* \\ &= cQ'(z,z). \end{aligned}$$

In fact, as we shall prove below, this implies the equation

(3.32) $$E_w^* = cE_w'^* \qquad w \in G/K.$$

But now, we're through. Let $f \in H$, and orthogonal to H′. For $w \in G/K$ and $\phi \in L_w$,

$$\langle f(w), \phi \rangle_{L_w} = \langle E_w f, \phi \rangle_{L_w} = \langle f, E_w^* \phi \rangle_H = c\langle f, E' \phi_w^* \rangle_H = 0$$

since $E_w'^* \phi \in H'$. Thus $f(w) = 0$ for all w, so $f \equiv 0$.

3.32. **Lemma.** *Let Z be a connected complex manifold and Z* the complex manifold obtained from Z by using the opposite complex structure. Let Δ be the diagonal in Z×Z*. Then any section of a holomorphic bundle V over Z×Z* which vanishes on Δ vanishes identically.*

Proof. The lemma is local, so we may assume $Z \subset \mathbb{C}^n$ and take $Z^* = \{\bar{z};\ z \in \mathbb{C}^n.\ \Delta \subset Z \times Z^*\}$ is the set of (z, w) such that $w = z$. Let $u = i(z-w)$, $v = z+w$. Then u, v are holomorphic coordinates on $Z \times Z^*$, and Δ is the set where u, v are real. The lemma now follows since any holomorphic function which vanishes when the coordinates are real must vanish identically.

Now, we can take $Z = G/K$ and Z^* as G/K with the opposite complex structure. Let V be the vector bundle over $Z \times Z^*$ whose fiber at (z, w) is $L_z \times L_w^* \cong \mathrm{Hom}(L_w, L_z)$. Then Q, Q' are holomorphic sections of V and $Q - cQ'$ is zero on the diagonal. Thus

$$Q(z, w) = cQ'(z, w) \qquad z, w \in K$$

Since $E^*(z) = Q(z, w)$, (3.32) follows.

3.33. Representations in cohomology

Let G be a semi-simple Lie group with maximal torus T contained in the maximal compact group K. Since root spaces are one-dimensional, for every group $H \supset T$, $h^{\mathbb{C}}$ is a sum of root spaces. Thus, if Δ^+ is a choice of positive roots, we let Δ_c^+ be those whose rootspaces are in $k^{\mathbb{C}}$, and Δ_{nc}^+ the others. Then the decompositions

$$g^{\mathbb{C}} = t^{\mathbb{C}} \oplus g^+ \oplus g^-$$

$$k^{\mathbb{C}} = t^{\mathbb{C}} \oplus g_c^+ \oplus g_c^-$$

exhibit G/T and K/T as complex manifolds. Writing

$$g^{\mathbb{C}} = k^{\mathbb{C}} \oplus g_{nc}^+ \oplus g_{nc}^-,$$

we have the realization of G/K as a complex manifold and the fibration

$$K/T \xrightarrow{\ i\ } G/T \xrightarrow{\ \pi\ } G/K$$

is holomorphic. However, if the choice of positive root system changes in one of these representations, the fibration need not be holomorphic. More generally, we may take H as a subgroup of K containing T so that each of the terms in the fibration

$$K/H \xrightarrow{\ i\ } G/H \xrightarrow{\ \pi\ } G/K$$

is a complex manifold, and so that i is holomorphic but π is not. This situation arises in the period domains of Griffiths (such as G/H), and was studied by Wells and Wolf [14]. The situation is this: G/H is s-concave, where $s = \dim_C K/H$. Thus we anticipate finding infinite-dimensional cohomology in $H^s(G/H, V)$ where V is a holomorphic vector bundle on G/H. Let $\Lambda = R\pi_s^* V$ be the sth direct image of V under π. Vaguely speaking, Λ is the vector bundle on G/K whose stalk at any point z is $H^s(\pi^{-1}(z), V)$. Since every vector bundle on G/K has a holomorphic structure, we may consider $\Lambda \longrightarrow G/K$ a holomorphic vector bundle. Since G/K is a Stein manifold, $H^0(G/K, \Lambda)$ is infinite-dimensional. Now, sections of Λ lift back to cohomology classes in V (this is a central theorem), so $H^s(G(H, V)$ is infinite-dimensional. There is a subspace of $H^0(G/K, \Lambda)$ (maybe $\{0\}$) which can be given a G-invariant Hilbert space structure, which pulls back to $H^s(G/H, V)$.

The case of SU(2, 2) considered above is slightly more general: in that case H is a subgroup of another subgroup K′ of SU(2, 2), and K′ is not compact, but the fibration

$$K'/H \xrightarrow{i'} G/H \xrightarrow{\pi'} G/K'$$

is holomorphic. The Penrose correspondence lifts cohomology on G/K′ back to G/H and then uses the pushforward above. Here $K = S(U(2)\times U(2))$, $H = S(U(1)\times U(1)\times U(2))$ and $K' = S(U(1)\times U(1,2))$. In the final section we shall extend this picture to subgroups of Sp(n, R).

4. The Heisenberg and Symplectic Groups.

Let $V = R^n \oplus R^{n*}$, where the second vector is considered as the dual of the first. Let $z = \begin{pmatrix} x \\ y \end{pmatrix}$ represent a point in V, and $\langle x, y\rangle$ the pairing between R^n and R^{n*}.

4.1. Definition. *The* symplectic form *on* V *is the nondegenerate skew* 2*-form* B*;*

$$B(z, z') = \langle x', y\rangle - \langle x, y'\rangle.$$

In the direct sum composition, B is represented by the block matrix

$$J = \begin{pmatrix} 0 & -I \\ I & 0 \end{pmatrix}: \ B(z, z') = {}^t z J z'$$

where I is the n×n identity matrix. We can now define the two groups.

4.2. Definition. *The* symplectic group*,* Sp(n, R)*, is the invariance group of* J*:*

(4.3) $$Sp(n, R) = \{g \in GL(V);\ {}^t g J g = J\}.$$

The vector space $V \oplus R$ can be made into a group by the group law

(4.4) $$(z, t)(z', t') = (z+z', t+t' + B(z, z')).$$

This is the *Heisenberg group* H_n: it is a two-step nilpotent group with center R.

Now the natural action of Sp(n, R) on V when extended trivially to $V \oplus R$, defines an automorphism of H_n leaving the center pointwise-fixed.

4.5. Definition. *Let* B_0 *be the group of automorphisms of* H_n *leaving the center pointwise-fixed.*

Then B_0 is in fact the semi-direct product of Sp(n, R) and H_n/R (acting as inner automorphisms). This group is central to the ideas of Weil in [13], (see also [4]), which is the foundation of the material to be described.

. The Lie algebra of H_n is again $V \oplus R$ with the bracket

(4.6) $$[(z, t), (z', t')] = (0, 2B(z, z')).$$

If we consider $z = x + iy$ as a complex variable (we give V the complex structure defined by J, since $J^2 = -I$), and let h be the subspace of $h_n^{\mathbb{C}}$ defined by such z, h is a Lie algebra of the type considered in section 3:

$$h_n^{\mathbb{C}} = \mathbb{C} \oplus h \oplus \bar{h},$$

where the first factor is the center. If $\chi((0, t)) = \exp 2\pi i t$, χ is a character of the center, and the theory of Section 3 leads to the representation known as the *Fock representation.* Skipping the details, we get:

(4.7) $$H = \{f \in O(\mathbb{C}^n);\ \int |f(z)|^2 e^{-2\pi \|z\|^2} dV < \infty\}.$$

Letting $F: H_n \longrightarrow U(H)$, we have

(4.8) $$(F(z_0, t_0) f)(z) = \exp \pi(-2it_0 + \|z_0\| + 2\langle z, z_0 \rangle) f(z - z_0).$$

For g an automorphism of H_n, conjugation by g gives a new representation F_g on $U(H)$:

$$F_g(n) = F(g \cdot n).$$

In particular, if g is a unitary transformation of $\mathbb{C}^n$, g is in $Sp(n, R)$ (under the isomorphism $C^n \cong (V, J)$), and thus defines an automorphism of H_n. g also acts unitarily on H by the contravariant action W:

$$(W(g) \cdot f)(z) = f(g^{-1} z).$$

It is readily verified that conjugation by $W(g)$ on $U(\mathcal{H})$ gives a realization of F_g:

(4.9) $$F_g(n) = W(g)^{-1} F(n) W(g).$$

It is easy to guess that this works for all automorphisms of H_n: this is almost true and gives what Weil calls the *"metaplectic representation".* It follows from the theorem of Stone and von Neumann on the uniqueness of the Heisenberg commutation relations. In our context this theorem says the following

4.10. **Theorem.** *Let* $R: H_n \longrightarrow U(V)$ *be a unitary representation of* H_n *on a Hilbert space* V *so that* $R((0, t))$ *is given by multiplication by the scalar* $\exp(2\pi i t)$. *Thus, there is a unitary transformation* $W: V \longrightarrow H$, *called an* intertwining operator *such that*

$$R(n) = W^{-1} F(n) W.$$

W is determined up to a scalar of modulus one. Thus, for any $g \in Sp(n, R)$, there is a $W(g) \in U(H)$ such that (4.9) holds. $W(g)$ is determined up to a scalar; this gives what is called a *projective representation* of $Sp(n,R)$. The scalar can be continuously determined on the two-sheeted cover $Mp(n, R)$ of $Sp(n, R)$, giving a representation (the "Weil", "metaplectic" or "harmonic oscillator" representation) of $Mp(n, R)$.

Now we bring in the last ingredient: we can polarize h_n by picking any complex structure on V which is compatible with the symplectic structure.

4.12. Definition.

$$S = \{\beta \in GL(V);\ \beta^2 = -I,\ {}^t\beta J \beta = J\}$$
$$= \{\beta \in Sp(n, R);\ \beta^2 = -I\}.$$

$Sp(n, R)$ acts on S by conjugation $\beta \longrightarrow g^{-1}\beta g$. For $\beta \in S$, $H_\beta = J\beta + iJ$ is a hermitian form on the complex vector space (V, β), and β is its imaginary part.

4.13. Definition. $S_p = \{\beta \in S;\ H_\beta$ has signature $(p, n-p)\}$: H_β *has* p *positive eigenvalues and* $n-p$ *negative eigenvalues.*

Now, up to complex isomorphism, there is only one vector space with hermitian form of signature $(p, n-p)$, and its invariance group is (by definition) $U(p, n-p)$. This translates to

4.14. Proposition. $Sp(n, R)$ *acts transitively on* S_p *and, as homogeneous spaces,*

$$S_p \cong Sp(n, R)/U(p, n-p).$$

Each element $\beta \in S$ thus defines a polarization on h_n: we seek the corresponding Fock space. For $\beta \in S_n$, this is easy: just replace $\langle z, w\rangle$ by $H_\beta(z, w)$ in (4.7) and (4.8). If $\beta \in S_p$, $p < n$, there are no holomorphic functions, and the representation exists in cohomology [3, 8].

More precisely, for $\beta \in S_p$, let $\mu_1^2, \dots, \mu_p^2, -\mu_{p+1}^2, \dots, -\mu_n^2$ be the eigenvalues of H_β, and let $z_1, \dots, z_n$ be coordinates dual to orthonormal basis of eigenvectors. Let

$$v_j = \mu_j z_j, \qquad j \leqslant p, \qquad v_k = \mu_k \bar{z}_k, \qquad h > p.$$

Then the cohomology space corresponding to β is given by

$$H_\beta = \{\omega = f(v)\exp(-\pi \|v\|^2)\, d\bar{z}_{p+1} \wedge \dots \wedge d\bar{z}_n;$$

$$f \in O(n)$$

$$\|\omega\|^2 = \frac{1}{(\mu_1 \dots \mu_n)^2} \int |f|^2 e^{-\pi \|u\|^2} dV < \infty\}.$$

The action of H_n is given by (4.8) in the v-coordinates.

According to the Stone-von Neumann theorem there are intertwining operators among all these representations and in fact the coordinate choice above describes these operators. The metaplectic representation can also be described in each of these spaces, and in doing this on S_n we shall relate this to the discrete holomorphic series of Sp(n, R) (representations on Sp(n, R)/U(n)), [16]. When we succeed in doing the same for the S_p, $p<n$, hopefully shall we find the representations on cohomology spaces on Sp(n, R)/U(p, q). We shall say more about this later. For now, we describe the situation for S_n.

For $\beta \in S$, we get a splitting

$$V^{\mathbb{C}} = h_\beta \oplus \overline{h}_\beta,$$

and the correspondence $\beta \longleftrightarrow h_\beta$ realizes S as

$$\{h \in \mathrm{Gr}_{\mathbb{C}}(2n, n);\ h \text{ is Lagrangian},\ h \cap \overline{h} = \{0\}\}.$$

The S_p are the open orbits of Sp(n, R) and thus inherit a homogeneous complex structure from $\mathrm{Gr}_{\mathbb{C}}(2n, n)$. The splitting $V = R^n \oplus R^{n*}$ lifts to a splitting $V^{\mathbb{C}} = \mathbb{C}^n \oplus \mathbb{C}^n$ so that both factors are Lagrangian. This splitting determines a coordinate neighborhood of $\mathrm{Gr}_{\mathbb{C}}(2n, n)$ biholomorphic to $\mathrm{End}(\mathbb{C}^n)$:

(4.16) $$\tau \longleftrightarrow \left\{\begin{pmatrix} \tau v \\ v \end{pmatrix};\ v \in \mathbb{C}^n\right\},\quad \tau \in \mathrm{End}(\mathbb{C}^n).$$

In this coordinatization S corresponds to symmetric matrices, with $\mathrm{Im}\,\tau$ giving the hermitian form $J\beta + iJ$. In particular S_n is completely contained in this coordinate neighborhood, and

(4.17) $$S_n \cong \{\tau \in \mathbb{C}^{n\times n};\ {}^t\tau = \tau,\ \mathrm{Im}\,\tau >> 0\},$$

the Siegel upper-half plane.

Now, if $V \longrightarrow S$ is the "tautological bundle", there is an induced action of Sp(n, R). Since S_n is contained in this coordinate neighborhood, we can trivialize this action: we have a map $S_n \times \mathbb{C}^n \longrightarrow V|S_n$ given by

$$(\tau, z) \longrightarrow \left(\tau, \begin{pmatrix} \tau z \\ z \end{pmatrix}\right),$$

and the Sp(n, R) action is given by

(4.18) $$g(\tau, z) = ((a\tau+b)(c\tau+d)^{-1}, (c\tau+d)^{-1})$$

for $g = \begin{pmatrix} a & b \\ c & d \end{pmatrix} \in \mathrm{Sp}(n, R)$.

Now, we can incorporate the H_n-action by recalling that a point $\tau \in S$ corresponds to a polarization of the Heisenberg Lie algebra, which is a line bundle over the fiber V_τ of V. Thus, we have a fiber-preserving action of H_n on $S_n \times \mathbb{C}^n \times \mathbb{C}$ given by

$$(4.19) \qquad n(\tau, z, w) = (\tau, z+z_0, w\exp(2\pi i t_0 - \pi\, {}^t\bar{z}_0 (\operatorname{Im}\tau) z),$$

where $\qquad n = (z_0, t_0) \in H_n$.

Formulae (4.18), (4.19) describe the action of B_0 on $S_n \times \mathbb{C}^n \times \mathbb{C}$ considered as a line bundle over $S_n \times \mathbb{C}^n$. The space of sections of this line bundle is the space of theta-functions, Θ. More precisely, we ask that they lie in the space H_β on the fiber over β. The contravariant action of B_0 on Θ is related both to the metaplectic representation and to the holomorphic series of representations of Sp(n, R). The theta-functions classically are the elements of Θ which are invariant under the action of the subgroup of H_n of elements with integer entries. This space is also invariant under Sp(n, Z) and the expression of this invariance in the above coordinates gives the classical conditions defining the theta-functions.

Now, the Schrodinger representation of H_n is the representation on $L^2(R^n)$ given by

$$S(x, 0, 0))f(x) = f(u - x_0)$$

$$S((0, y, 0))f(x) = \exp(2\pi i\langle u, y_0\rangle f(u)$$

$$S((0, 0, t))f(x) = \exp(2\pi i t) f(u).$$

By the Stone-von Neumann theorem there is an intertwining operator $I_\tau \colon L^2(R^n) \longrightarrow \mathcal{H}_\tau$; a calculation shows that we can take

$$(I_\tau f)(z) = e^{-z\tau z} \int_{R^n} e^{i\,{}^t u z + \frac{i}{2}\,{}^t u \tau u} f(u)\, du.$$

Notice that I_τ is holomorphic in τ, so we can think of $I \colon L^2(R^n) \longrightarrow \Theta$

$$(I f)(\tau, z, w) = (I_\tau f)(z) w.$$

This I intertwines the two representations of B_0, and we anticipate that similar intertwining operators will work for the other S_p, $p < n$.

Fix a $p < n$, and let

$$E = \{(S, \beta);\ \beta \in S_n,\ S \text{ is a p-dimensional subspace of } h_\beta\},$$

$$E' = \{(s, \beta);\ \beta \in S_p,\ S \text{ is a p-dimensional subspace of } h_\beta \text{ and } H_\beta | S >> 0\}.$$

E and E′ are in one-one correspondence as follows: for $(S, \beta) \in E$, let S' be the subspace of h_β orthogonal to S. Then $S \oplus S' = h_{\beta'}$, for some $\beta' \in S_p$ (recall (4.15)). This map $(S, \beta) \longrightarrow (S, \beta')$ defines the correspondence, and commutes with the Sp(n, R)-action. In fact, both E and E′ are Sp(n, R)-homogeneous and in both cases the isotropy group is (conjugate to) U(p)×U(q). We thus get a "generalized Penrose correspondence"

$$\begin{array}{ccc} & P & \\ E & \rightarrow & E' \\ \tau_p \downarrow & & \downarrow \tau_n \\ \mathcal{S}_p & & \mathcal{S}_n \end{array} \tag{4.20}$$

where the vertical arrows are given by the projection on the second factor. $\mathcal{S}_n$ parametrizes some of the compact subvarieties of $\mathcal{S}_p$ of dimension p(n−p), and since $\mathcal{S}_p$ has p(n−p)-concavity, we anticipate a correspondence between $H^{p(n-p)}$-spaces on $\mathcal{S}_p$ and holomorphic functions on $\mathcal{S}_n$. Unfortunately the correspondence at the top does not preserve the complex structure, so this is not evidently the case.

REFERENCES

[1]. A. Andreotti and F. Norguet, *Problème de Levi et Convexité holomorphe pour les classes de cohomologie,* **Ann. Scuola Norm. Pisa Cl. Sci.** 20 (1966), 197-241.

[2]. L. Auslander and B. Kostant, *Polarizations and Unitary Representations of Solvable Lie Groups,* **Invent. Math** 14 (1971), 255-354.

[3]. J. Carmona, *Representations du groupe de Heisenberg dans les espaces de* (0, q) *formes,* **Math. Annalen,** 205 (1973), 89- 112.

[4]. J. Igusa, *Theta Functions,* **Grund. Math. Wiss.** vol 194, Springer Verlag, (1972).

[5]. B. Kostant, *Quantization and Unitary Representations,* **Lecture Notes in Math.** vol. 170, Springer Verlag, 1970

[6]. R. Kunze, *On the Irreducibility of Certain Multiplier Representations,* **B. A. M. S.** 68 (1962), 93-94.

[7]. H. Rossi and M. Vergne, *Representations of Certain Solvable Lie Groups on Hilbert Spaces of Holomorphic Functions and the Application to the Holomorphic Discrete Series of a Semisimple Lie Group,* **J Funct. Anal.** 13 (4) (1973), 324-389.

[8] H. Rossi and M. Vergne, *Group Representations on Hilbert Spaces Defined in Terms of d_b-cohomology on the Silov Boundary of a Siegel Domain,* **Pacific J Math.** 65 (1976), 193-207.

[9]. H. Rossi and M. Vergne, *Equations de Cauchy-Riemann tangentielles asociées à un domaine de Siegel,* **Ann. Scient. E. N. S.** vol. 9 (1976), 31-80

[10]. H. Rossi and M. Vergne, *Analytic Continuation of the Holomorphic Discrete Series of a Semi-simple Lie Group,* **Acta Math.** 136 (1976), 1-59.

[11]. I. Satake, *Factors of Automorphy and Fock Representations,* **Advances in Mathematics** 7 (1971), 83-111.

[12]. R. Tolimieri, *Heisenberg Manifolds and Theta Functions,* **T. A. M. S.** 239 (1978), 293-319.

[13]. A. Weil, *Sur certaines groupes d'operateurs unitaires,* **Acta Math.** 111 (1964), 143-211.

[14]. R. O. Wells and J. A. Wolf, *Poincaré Series in Automorphic Cohomology on Flag Domains,* **Ann. of Math.** 105 (1977), 297-448.

[15]. R. O. Wells, *Complex Manifolds and Mathematical Physics,* **B. A. M. S. (N. S.)** 1 (1979), 196-336.

[16]. N. Kashiwara and M. Vergne, *On the Segal-Shale-Weil Representations and Harmonic Polynomials,* **Inv. Math.** 44 (1978), 1-47.

School of Mathematics
The Institute for Advanced Study
Princeton, N.J. 08540
U.S.A

B SCHIFFMAN* & A J SOMMESE

Vanishing theorems for weakly positive vector bundles

Introduction.

While the Kodaira vanishing theorem generalizes to weakly positive line bundles, the same is not true for the Nakano vanishing theorem. However, some extensions of the Nakano vanishing theorem to weakly positive line bundles are possible; one such generalization is due to Girbau [2]. In this article we give some other generalizations (theorems (2.8) and (2.9)) in which we assume that the hypotheses of the Nakano or Girbau vanishing theorems are satisfied off a subvariety. We also define the concept of k-positivity for vector-bundles and use a theorem of Le Potier [6] to state our result for vector bundles. Some similar generalizations to k-ample line bundles are given in [9] and in [8, Chapter III].

The first section reviews our notation and summarizes some basic vanishing theorems. For a more detailed discussion of the facts from Section I and of vanishing theorems in general, the reader may consult [8].

1. Vanishing theorems of Kodaira, Nakano, Girbau and Le Potier.

Let E be a hermitian holomorphic vector bundle on an n-dimensional compact Kähler manifold X. We let $C^\infty(X, E)$ denote the space of C^∞-sections of E over X. We let $T = T_X$ and $T^* = T^*_X$ denote the holomorphic tangent and cotangent bundles, respectively, of X. We give E the hermitian connection $\nabla = \nabla' + \nabla''$ where

(1.1)
$$\nabla' : C^\infty(X, E) \longrightarrow C^\infty(X, T^* \otimes E).$$
$$\nabla'' = \bar\partial : C^\infty(X, E) \longrightarrow C^\infty(X, \bar T^* \otimes E),$$

where $\bar T^*$ is the anti-holomorphic bundle of (0, 1)-forms. The hermitian connection ∇ is characterized by the conditions that it preserves the metric on E and that $\nabla'' = \bar\partial$. The curvature form

(1.2)
$$\Theta \in C^\infty(X, \mathrm{Hom}(E, E) \otimes T^* \otimes \bar T^*)$$

is given by $\nabla^2 s = \Theta s$, for $s \in C^\infty(X, E)$. Let

(1.3)
$$E^{p,q} = \Lambda^p T^* \otimes \Lambda^q \bar T^* \otimes E$$

denote the bundle of E-valued (p, q)-forms. Then we have

*Research Supported in part by the National Science Foundation

$$\nabla': C^\infty(X, E^{p,q}) \longrightarrow C^\infty(X, E^{p+1,q}),$$

(1.4)

$$\nabla'': C^\infty(X, E^{p,q}) \longrightarrow C^\infty(X, E^{p,q+1}).$$

We let $\delta' = -*\nabla''*$, $\delta'' = -*\nabla'*$, where

(1.5) $$*: E^{p,q} \longrightarrow E^{n-q,n-p}$$

denotes the Hodge $*$-operator. We define the Laplacians

$$\Delta' = \delta'\nabla' + \nabla'\delta'$$

(1.6)

$$\Box = \Delta'' = \delta''\nabla'' + \nabla''\delta''.$$

Let $\Omega^p(E)$ denote the sheaf of germs of holomorphic p-forms on E. Thus $\Omega^n(E) = \mathcal{O}(K_X \otimes E)$, where $K_X = \Lambda^n T_X$. The basic result of complex Hodge theory states that

(1.7) $$H^q(X, \Omega^p(E)) \approx \{\alpha \in C^\infty(X, E^{p,q}): \Box\alpha = 0\},$$

where the isomorphism is naturally given in terms of Dolbeault cohomology. We let $L: E^{p,q} \longrightarrow E^{p+1,q+1}$ be given by $L\alpha = \omega \wedge \alpha$, where ω is the Kähler form of X, and we let

(1.8) $$\Lambda = L^* = (-1)^{p+q} * L *: E^{p,q} \longrightarrow E^{p-1,q-1}.$$

We have the *Kodaira-Nakano Identity*

(1.9) $$\Box = \Delta' + \sqrt{-1}(\nabla^2\Lambda - \Lambda\nabla^2),$$

which depends essentially on X being Kähler.

We introduce the following definition, which we shall extend to vector bundles in Section 2:

(1.10) **Definition.** *Let* E *be a hermitian holomorphic line bundle on an n-dimensional complex manifold* X, *and let* $\Theta \in C^\infty(X, T^* \otimes \overline{T}^*)$ *be the curvature form of* E. *We say that* E *is* k-positive *at* $x \in X$ *if the hermitian form* Θ_x *is positive semi-definite and has at least* n−k *positive eigenvalues on* T_x. *A holomorphic line bundle* E *is said to be* k-positive *if* E *carries a hermitian metric that is k-positive at all points of* x.

Recall that a line bundle on a compact complex manifold is *positive* (or *ample*) if it is 0-positive. The following vanishing theorem of Akizuki and Nakano [1] is a consequence of (1.7) and (1.9):

(1.11) **Nakano Vanishing Theorem.** *Let* E *be a positive holomorphic line bundle on a compact complex manifold* X. *Then*

$$H^q(X, \Omega^p(E)) = 0 \qquad \text{for } p+q > \dim X.$$

Note that the hypotheses of (1.11) imply the Kodaira embedding theorem [5] that X is projective-algebraic. The case $p = \dim X$ of (1.11) is due to Kodaira [5]. However, this vanishing theorem of Kodaira is valid with the following weaker hypotheses:

(1.12) **Kodaira Vanishing Theorem.** *Let* E *be a holomorphic line bundle on a compact Kähler manifold* X. *Suppose* E *has a hermitian metric with curvature form that is positive semi-definite on all of* X *and* k*-positive at some point of* X. *Then*

$$H^q(X, O(K_X \otimes E)) = 0 \qquad \text{for } q > k.$$

The Nakano Vanishing Theorem (1.11) cannot be extended to line bundles that are positive on only part of X. For example, let $X \subset \mathbb{P}^3 \times \mathbb{P}^2$ be the 3-manifold obtained by blowing up a point x_0 of $\mathbb{P}^3$. Let $\pi_1: X \longrightarrow \mathbb{P}^3$, $\pi_2: X \longrightarrow \mathbb{P}^2$ be the projections, and let $S = \pi_1^{-1}(x_0)$. Let $E = \pi_1^*(H)$, where H is the hyperplane section bundle on $\mathbb{P}^3$. Then E has a metric that is semi-positive on X and positive on X–S. However, $H^2(X, \Omega^2(E)) \neq 0$, since $\pi_2^*(\omega^2) \otimes \sigma$ is not cohomologous to 0, where ω is the Kähler form on $\mathbb{P}^2$ and σ is a section in $\Gamma(X, O(E))$ such that $\sigma|S \neq 0$.

However, the Nakano Vanishing Theorem (1.11) does generalize to k-positive line bundles:

(1.13) **Theorem** *(Girbau* [2]*). If* E *is a* k*-positive holomorphic line bundle on a compact Kähler manifold* X, *then*

$$H^q(X, \Omega^p(E)) = 0 \qquad \text{for } p+q > n+k.$$

To generalize these vanishing theorems to vector bundles, we need the following construction of Grothendieck [4]: Let E be a holomorphic vector bundle of rank $r > 1$ on a complex manifold X. We define the holomorphic fibre bundle

(1.14) $$\pi: P(E) \longrightarrow X,$$

where the fibre $\pi^{-1}(x)$ is the space of (r–1)-dimensional subspaces of E_x, for $x \in X$. (Thus $\pi^{-1}(x)$ is naturally isomorphic to the projective space of lines in E_x^*.) The space P(E) carries a tautological subbundle F of $\pi_* E$ given by

(1.15) $$F_y = y \subset (\pi^* E)_y = E_{\pi(y)}, \qquad \text{for } y \in P(E).$$

Note that rank $F = r-1$. We define the holomorphic line bundle ξ_E on P(E) by

(1.16) $$\xi_E = \pi^* E/F.$$

The bundle $\xi_E \longrightarrow P(E)$ is called the line bundle associated to E. We shall use the following result:

(1.17) **Theorem** *(Le Potier* [6]*). Let* E *be a holomorphic vector bundle on a complex manifold* X *and let* $\mathcal{S}$ *be a coherent analytic sheaf on* X. *Then*

$$H^q(X, \mathcal{S} \otimes \Omega^p_X(E)) = H^q(P(E), \pi^* \mathcal{S} \otimes \Omega^p_{P(\hat{E})}(\xi_E))$$

for all $p, q \geqslant 0$.

A holomorphic vector bundle E on a compact complex manifold X is said to be *ample* if ξ_E is a positive line bundle on $P(E)$. The following result is an immediate consequence of Theorems (1.11) and (1.17):

(1.18) **Corollary** *(Le Potier Vanishing Theorem* [6]*). If E is an ample vector bundle on a compact complex manifold X, then*

$$H^q(X, \Omega^p(E)) = 0 \qquad \text{for} \quad p+q \geqslant \dim X + \operatorname{rank} E.$$

As in (1.11), the hypotheses of corollary (1.18) imply that X is projective-algebraic.

2. Generalizations of the Girbau and Le Potier vanishing theorems.

We first extend the concept of k-positivity to vector bundles. Let E be a holomorphic vector bundle on a compact complex manifold X. Suppose E is given a hermitian metric H. Recalling (1.2), we have the curvature form

$$\Theta \in C^\infty(X, E^* \otimes E \otimes T^* \otimes \overline{T}^*)$$

where we identify $\operatorname{Hom}(E, E)$ with $E^* \otimes E$. Let $\hat{H}: E \longrightarrow \overline{E}^*$ denote the C^∞-vector bundle isomorphism induced by H. We define the curvature tensor

(2.1) $$\mathcal{R} = \hat{H}\Theta \in C^\infty(X, E^* \otimes \overline{E}^* \otimes T^* \otimes \overline{T}^*).$$

Let $\{\epsilon^1, \ldots, \epsilon^r\}$ be a local frame for E^*, and let $h = (h_{ij})$ be the matrix for H with respect to $\{\epsilon^i\}$; that is

$$H = \Sigma h_{ij}\, \epsilon^i \otimes \overline{\epsilon}^j.$$

Then

(2.2) $$\mathcal{R} = \Sigma\, \epsilon^i \otimes \overline{\epsilon}^j \otimes (-\partial\overline{\partial} h + \partial h \wedge h^{-1} \overline{\partial} h)_{ij}.$$

In particular $\mathcal{R}_x$ can be regarded as a hermitian form on $(E \otimes T)_x$. The strictest form of positivity for a vector bundle is Nakano-positivity. We say that E is *Nakano-positive* at x if $\mathcal{R}_x$ is a positive-definite hermitian form on $(E \otimes T)_x$; E is *Nakano semi-positive* at x if $\mathcal{R}_x$ is semi-positive on $(E \otimes T)_x$. The following vanishing theorem is due to Nakano [7].

(2.3) **Theorem.** *Let E be a hermitian holomorphic vector bundle on a compact Kähler manifold X. If E is Nakano-semi-positive at all points of X and Nakano-positive at $x_0 \in X$, then*

$$H^q(X, \mathcal{O}(K_X \otimes E)) = 0 \qquad \text{for} \quad q \neq 0.$$

Nakano-positivity is a difficult condition to verify. In fact T_{P^n} is not Nakano-positive, although it is ample. We now give the concept of k-positivity for vector bundles, based on Griffith's concept of positivity:

(2.4) **Definition.** *Let* E *be a hermitian holomorphic vector bundle on an* n*-dimensional complex manifold* X. *We say that* E *is* k-positive *at* $x \in X$ $(0 \leqslant k \leqslant n-1)$ *if for all non-zero* $e \in E_x$, *the hermitian form*

$$\mathcal{R}_x \cdot (e \otimes \overline{e}) \in T_x^* \otimes \overline{T}_x^*$$

is semi-positive and has at least $n-k$ *positive eigenvalues on* T_x.

A holomorphic vector bundle E on a compact complex manifold X is said to be *k-positive* if E has a hermitian metric that is k-positive at all points of X. We note that 0-positive corresponds to Griffiths-positive [3]. We also have the implications

$$\text{Nakano-positive} \Longrightarrow \text{0-positive} \Longrightarrow \text{ample}$$

The first of the above implications is a tautology; the second follows from lemma (2.6) below. We note the following generalization of the Le Potier Theorem:

(2.5) **Theorem** *(Girbau* [2]*). Let* E *be a* k*-positive holomorphic vector bundle on a compact Kähler manifold* X. *Then*

$$H^q(X, \Omega^p(E)) = 0 \quad \text{for} \quad p + q \geqslant k + \dim X + \operatorname{rank} E.$$

We give a proof of theorem (2.5), using the following two lemmas:

(2.6) **Lemma.** *If* E *is* k*-positive, then* ξ_E *is a* k*-positive line bundle on* P(E).

Proof. Give E a k-positive hermitian metric, and give ξ_E the quotient metric induced from the pull-back metric on π^*E. Let $\mathcal{R}$ and $\mathcal{R}'$ denote the curvature tensors of π^*E and ξ_E respectively. Let $y \in P(E)$ be arbitrary and choose a unit vector $e \in (\pi^*E)_y = E_{\pi(y)}$ such that e is orthogonal to y. Let e′ denote the image of e in ξ_E. Let

$$V = \ker \pi_{*y} \subset T_{P(E),y}$$

denote the vertical tangent space at y. By hypothesis the hermitian form $\mathcal{R}_y \cdot (e \otimes \overline{e})$ is semi-positive and has at least $n-k$ positive eigenvalues on $T_{P(E),y}$; let W denote its positive eigenspace. Thus $\dim W \geqslant n-k$ and furthermore $V \cap W = (0)$. Since

$$\mathcal{R}'_y = \mathcal{R}'_y \cdot (e' \otimes \overline{e}') \geqslant \mathcal{R}_y \cdot (e \otimes e)$$

by the increasing property of curvature under quotient maps, and since $\xi_E|_{\pi^{-1}(x)}$ is the hyperplane-section bundle with positive curvature, it follows that $\mathcal{R}'_y$ is semipositive on $T_{P(E),y}$ and positive definite on $V \oplus W$. Since $\dim V \oplus W \geqslant \dim P(E) - k$, this completes the proof of the lemma.

(2.7) **Lemma.** *If* E *is a holomorphic vector bundle on a compact Kähler manifold* X, *then* $P(E)$ *is Kähler.*

Proof. Let ω be the Kähler form on X. Identify $P(E)$ with the bundle of lines in E^*, and give E^* a hermitian metric. Let $\phi: E^* \longrightarrow \mathbb{R}$ be given by $\phi(\lambda) = ||\lambda||^2$ and let $\eta = \sqrt{-1}\,\partial\bar{\partial}\log\phi$, regarded as a (1, 1)-form on $P(E)$. Then $\pi^*\omega + \epsilon\eta$ is a Kähler form on $P(E)$ for sufficiently small positive ϵ.

Proof of Theorem (2.5): Let $E \longrightarrow X$ be as in the statement of the theorem. By Lemmas (2.6) and (2.7) ξ_E is a k-positive line bundle on a compact Kähler manifold. The conclusion follows from Girbau's Theorem (1.13) and Le Potier's Theorem (1.17).

We now give extensions of the Nakano and Girbau vanishing theorems to weakly positive bundles and vector bundles:

(2.8) **Theorem.** *Let* E *be a holomorphic vector bundle on a projective-algebraic manifold* X. *Suppose there exists a subvariety* A *of* X *and a hermitian metric on* E *such that* E *is Griffiths-positive at all points of* $X - A$. *Then*

$$H^q(X, \Omega^p(E)) = 0 \quad \text{for} \quad p+q \geqslant \dim X + \dim A + \operatorname{rank} E.$$

(2.9) **Theorem.** *Let* $E \longrightarrow X$ *be as in Theorem (2.8). Suppose there exists a subvariety* A *of* X *and a hermitian metric on* E *such that* E *is k-positive at all points of* $X - A$. *Then*

$$H^q(X, \Omega^p(E)) = 0 \quad \text{for} \quad p+q \geqslant \dim X + \dim A + \operatorname{rank} E + k + 1.$$

Recall that Griffiths-positive means 0-positive. Theorem (2.8) gives the vanishing of one more cohomology group (for each p) than Theorem (2.9) with $k = 0$. We do not know if Theorem (2.9) can be likewise improved for $k > 0$. The proof of Theorem (2.8) and (2.9) is based on the "slicing lemma" below.

We use the following notation: For an analytic hypersurface $D \subset X$, we let $[D]$ denote the holomorphic line bundle on X which has a global section with divisor equal to D.

(2.10) **Lemma.** *Let* E *be a holomorphic vector bundle on a complex manifold* X, *and let* D *be a smooth analytic hypersurface in* X. *Fix* $p, q \geqslant 0$. *If*

a) $H^q(X, \Omega^p_X([D] \otimes E)) = 0$,

b) $H^{q-1}(D, \Omega^{p-1}_D(E|_D)) = 0$,

c) $H^{q-1}(D, \Omega^p_D([D] \otimes E)|_D) = 0$,

then $H^q(X, \Omega^p(E)) = 0$.

Proof. Let $s \in \Gamma(X, [D])$ such that $\mathrm{Div}\, s = D$. Consider the short exact sequence of vector bundles on D,

(2.11)

$$0 \longrightarrow \Lambda^{p-1} T_D^* \otimes E|_D \xrightarrow{\wedge ds} (\Lambda^p T_X^* \otimes [D] \otimes E)|_D \longrightarrow \Lambda^p T_D^* \otimes ([D] \otimes E)|_D \longrightarrow 0.$$

By (b), (c) and the long exact cohomology sequence associated with (2.11), we conclude that

$$H^{q-1}(D, (\Lambda^p T_X^* \otimes [D] \otimes E)|_D) = 0. \qquad (2.12)$$

Consider the short exact sequence of sheaves on X,

$$0 \longrightarrow \Omega_X^p(E) \xrightarrow{\cdot s} \Omega_X^p([D] \otimes E) \longrightarrow \mathcal{O}_D \otimes \Omega_X^p([D] \otimes E) \longrightarrow 0. \qquad (2.13)$$

The desired vanishing follows from (a), (2.12) and the long exact cohomology sequence associated with (2.13). Q.E.D.

We now prove Theorem (2.9). Let $n = \dim X$, $r = \mathrm{rank}\, E$. We use induction on $\ell = \dim A$, where we define $\dim \phi = -1$. We note that for $\ell = -1$, Theorem (2.9) reduces to theorem (2.5). Let $\ell \geq 0$ and assume the theorem has been verified for $\dim A < \ell$. By Bertini's Theorem, we can choose a smooth hyperplane section $D \subset X$ such that $\dim A \cap D = \ell - 1$. Since the line bundle $E|_D$ is k-positive on $D - A \cap D$, it follows by induction that

$$H^q(D, \Omega_D^p(E|_D)) = 0 \qquad (2.14)$$

for $p+q \geq \dim D + \ell - 1 + r + k + l = n + k + r + \ell - 1$. Since [D] is positive, and E is k-positive, it follows that $[D] \otimes E$ is 0-positive. Thus by the Le Potier vanishing theorem (1.18),

$$H^q(X, \Omega^p([D] \otimes E)) = 0 \quad \text{for} \quad p+q \geq n+r. \qquad (2.15)$$

and

$$H^q(D, \Omega_D^p([D] \otimes E)|_D) = 0 \quad \text{for} \quad p+q \geq n-1+r. \qquad (2.16)$$

The desired vanishing theorem follows from (2.14)-(2.16) and Lemma (2.10).

The proof of Theorem (2.8) is exactly the same, except we must start the induction with $\dim A = 0$. However, in this case we can modify the metric on E so that it is 0-positive on all of X, as follows: Let h be the metric on E with 0-positive curvature on $X - A$, where $\dim A = 0$. Choose $\rho \in C^\infty(X)$ such that $\sqrt{-1}\, \partial \bar{\partial} \rho$ is positive at each point of A. One easily checks that the curvature form of the modified metric $h_\epsilon = e^{-\epsilon\rho} h$ is 0-positive on all of X for sufficiently small, positive ϵ. The conclusion follows by induction as before.

REFERENCES

[1]. Akizuki, Y., and Nakano, S., *Note on Kodaira-Spencer's Proof of Lefschetz Theorems,* **Proc. Jap. Acad.** 30 (1954), 266-272.

[2]. Girbau, J., *Sur le théorème de Le Potier d'Annulation de la cohomologie,* **C. R. Acad. Sci. Paris** 283 (1976), Serie A, 355-358.

[3]. Griffiths, P. A., *Hermitian Differential Geometry, Chern Classes, and Positive Vector Bundles,* in **Global Analysis, Papers in Honor of K. Kodaira,** Princeton Univ. Press, Princeton (1969), pp. 181-251.

[4]. Grothendieck, A., and Dieudonné, J., *Eléments de géométrie algébrique II,* **Publ. Math. I.H.E.S.** 8 (1961).

[5]. Kodaira, K., *On a Differential Geometric Method in the Theory of Analytic Stacks,* **Proc. Nat. Acad. Sci. U.S.A.** 39 (1953), 1268-1273.

[6]. Le Potier, J., *Annulation de la cohomologie à valeurs dans un fibré vectoriel holomorphe positif de rang quelconque,* **Math. Ann.** 218 (1975), 35-53.

[7]. Nakano, S., *On Complex Analytic Vector Bundles,* **J. Math. Soc. Japan** 7 (1955), 1-12.

[8]. Shiffman, B., and Sommese, A.J., *Vanishing Theorems on Complex Manifolds,* book to appear.

[9]. Sommese, A.J., *Submanifolds of Abelian Varieties,* **Math. Ann.** 233 (1978), 229-256.

The Johns Hopkins University
Baltimore, Maryland 21218
U.S.A.
University of Notre Dame
Notre Dame, Indiana 46556
U.S.A.

S M WEBSTER*

Real submanifolds of $\mathbb{C}^n$ and their complexifications

Introduction

The main purpose of this paper is to expound a basic philosophy toward the study of real submanifolds of $\mathbb{C}^n$. The first tenet of this philosophy is that one should first try to understand real analytic submanifolds and then use this understanding as a guide for the general case. The second is that real analytic submanifolds can best be understood by considering their complexifications. The appropriateness of this philosophy has been born out in a number of instances. To limit our discussion to a reasonable length we shall, aside from making some elementary remarks of a general nature, concentrate on the theory of local biholomorphic invariants. At present a systematic invariant theory has been found in only two cases. The first and best known is that of a real hypersurface with non-degenerate Levi form. The second is that of a real n-manifold in $\mathbb{C}^n$ near a suitable non-degenerate complex tangent.

The first to study a non-degenerate real hypersurface by complexification was B. Segre [10] in 1931. He observed that in $\mathbb{C}^2$ the complexification gives rise to an invariant two-parameter family of holomorphic curves. He then invoked the invariant theory of such curves, which are the solution curves of a second-order holomorphic ordinary differential equation. This theory had been worked out by A. Tresse [11] in 1896 in a prize essay. Shortly after Segre's work there appeared the two papers [3], in which E. Cartan gave a complete solution of the local problem via his famous "methode d'equivalence". After this the method of complexifications was not taken up again until the late 1970's.

In more than two complex variables one is led to an intrinsic family of complex hypersurfaces, which are solution manifolds to a differential system. The invariant theory of such a family was given in 1937 by M. Hachtroudi [7] using Cartan's method. This theory turns out to be the complexification of the Chern-Moser-Tanaka theory [5]. This relationship was considered by Chern [4] and by Faran in [6].

While one may to a great extent dispense with complexification in the study of non-degenerate real hypersurfaces, it seems to be absolutely essential to the understanding of a real analytic n-manifold in $\mathbb{C}^n$. The local theory of such has been developed by J. K. Moser and the present author in [9].

In this paper we shall present and attempt to compare the main ideas and results of these two theories. For the detailed proofs of these results we shall refer for the most part to the literature.

1. Basic Properties and Examples

We begin by showing how the basic concepts associated to an analytic real submanifold M may

* Partially supported by NSF, Grant No. MCS-8300245.

be motivated by its complexification. Let $z = x + iy$ be a coordinate vector for $\mathbb{C}^n$. If M has real codimension ℓ, then it is given locally as

(1.1) $$M:\ r = (r^1, \ldots, r^\ell) = 0, \qquad dr^1 \wedge \ldots \wedge dr^\ell \neq 0,$$

where $r = r(x, y) = r(z, \overline{z})$ is a real vector given by convergent power series. We replace $\overline{z}$ by an independent complex variable $\overline{w} = \eta$ to obtain the complexification,

(1.2) $$\mathfrak{M} = \{(z, \eta) \in \mathbb{C}^{2n}:\ r(z, n) = 0\}.$$

This manifests itself as the n-parameter family of local complex varieties

(1.3) $$Q_w = \{z \in \mathbb{C}^n:\ r(z, \overline{w}) = 0\}.$$

The function $r(z, \overline{w})$ is only defined on $\mathcal{U} \times \mathcal{U}$, where $\mathcal{U}$ is some neighborhood of $z_0 \in M$, and $Q_w \subset \mathcal{U}$; however, as in (1.3) we shall not bother to indicate this. Note that $z \in Q_w$ if and only if $w \in Q_z$ and that M is the set of points z such that $z \in Q_z$.

We may regard $M \subset \mathfrak{M}$ as the fixed point set of the anti-holomorphic involution $(z, \eta) \longrightarrow (\overline{\eta}, \overline{z})$. The utility of the complexification rests on the simple fact that if a function $h(z, \eta)$ holomorphic on $\mathfrak{M}$ vanishes on M, then it vanishes on all of $\mathfrak{M}$. This gives, for example, the biholomorphic invariance of the family Q_z. For, if $z' = f(z)$ maps M biholomorphically to M': $r' = 0$, then $r' \circ f$ vanishes on M: hence, $r'(f(z), \overline{f}(\overline{w}))$ vanishes on $\mathfrak{M}$, giving $f(Q_z) = Q_{f(z)}$.

If M: $r(z) = 0$ is a complex variety, then $M = Q_z$ for all z, and $\mathfrak{M} \cong M \times M$. The *complex envelope* of a general M may be constructed as follows. Let

(1.4) $$\pi_1(z, \eta) = z, \qquad \pi_2(z, \eta) = \eta$$

be the natural projections. Then $\pi_1(\mathfrak{M})$ is the smallest complex "variety" in $\mathbb{C}^n$ containing M. Since, if $f(z) = 0$ on M, then f (thought of as a function of (z, η)) must also vanish on $\mathfrak{M}$, i.e. $\pi_1(\mathfrak{M}) \subset f^{-1}(0)$. As an example, let $M^2 \subset \mathbb{C}^3$ be given by

(1.5) $$M: \begin{aligned} z_2 &= \overline{z}_2 = (z_1 + \overline{z}_1)^2 \\ z_3 &= \overline{z}_3 = (z_1 + \overline{z}_1)^3. \end{aligned}$$

Then $\pi_1(\mathfrak{M})$ is the singular variety $z_3^2 = z_2^3$. The surface $M^2 \subset \mathbb{C}^2$ given by $z_2 = \overline{z}_2 = z_1 \overline{z}_1$ shows that $\pi_1(\mathfrak{M})$ need not actually be a variety.

The tangent space $T_z Q_w : \partial_z r(z, \overline{w}) = 0$ coincides with the holomorphic tangent space $H_z(M)$ when $z = w \in M$. It follows that if M is a generic CR submanifold, i.e. $\partial r^1 \wedge \ldots \wedge \partial r^\ell \neq 0$, then each Q_w is nonsingular and of codimension ℓ. At a CR-singularuty of M (i.e. where $\dim_{\mathbb{C}} H_z(M)$ jumps) these properties may be lost. For example, let

(1.6) $$M:\ z_n = (\overline{z},\overline{z}), \qquad \overline{z}_n = (z,z) = \sum_{j=1}^{n-1} z_j z_j.$$

Then Q_w is singular for $w = 0$, while for

(1.7) $$M:\ z_n = \overline{z}_n = (z,\overline{z})$$

Q_w is smooth but jumps in dimension at $w = 0$. However, if M is a CR submanifold, then $\pi_1(\mathfrak{M})$ is a smooth complex submanifold and $M = \pi_1(M) \subset \pi_1(\mathfrak{M})$ is a generic embedding. In fact, $\partial r^1 \wedge \ldots \wedge \partial r^k \neq 0$ for some maximal $h \leqslant \ell$, and $\partial r^1 \wedge \ldots \wedge \partial r^k \wedge \partial r^\alpha = 0$ for any α. Since these relations persist on $\mathfrak{M}$, it is easily seen that $\pi_1\colon \mathfrak{M} \longrightarrow \mathbb{C}^n$ has constant rank. A study of submanifolds embedded generically away from their CR-singularities has been made by Harris [8].

The concepts of CR function, one annihilated by all tangential vectors of type (0.1), and of restriction of a holomorphic function are distinct in general. If $f = f(z)$ is holomorphic in a neighborhood of M in $\mathbb{C}^n$, then the analytic extension of $f|_M$ to $\mathfrak{M}$ must be constant on the fibers of π_1. These are of the form (z, Q_z), which (locally) are connected if M is generic CR. We may then assume $\det(\partial r^i/\partial z^j)_{1\leqslant i,j\leqslant \ell} \neq 0$. The tangent space to the fiber through (z, w), $w \in Q_z$, is spanned by the vectors

$$X_{\overline{\alpha}} = \det \left[\begin{array}{c|c} \partial_{\overline{\alpha}} & \partial_{\overline{j}} \\ \hline \partial_{\overline{\alpha}} r^i & \partial_{\overline{j}} r^i \end{array}\right]_{(z,\overline{w})}, \qquad \ell < \alpha \leqslant n,$$

(1.8)

$$(\partial_{\overline{j}} r = \partial r/\partial \overline{w}^j,\ \partial_j r = \partial r/\partial z^j, \text{ etc.})$$

where the determinant is expanded across the top row. The characterization $X_{\overline{\alpha}} f = 0$ reduces to the tangential Cauchy-Riemann equations on M. However, if M is the CR-singular surface

(1.9) $$M:\ z_2 = \overline{z}_2 = z_1\overline{z}_1 + \gamma(z_1^2 + \overline{z}_1^2), \qquad 0 < \gamma < \infty,$$

then the fiber of π_1 is generically a pair of points. The constancy of f on these fibers does not follow from the tangential Cauchy-Riemann equations, which restrict only the first derivatives of f at the origin, where M has an isolated complex tangent.

The concept of the Levi form in the CR case may be motivated as follows. The fibers (z, Q_z), (Q_w, w) of π_1, π_2, respectively, give two holomorphic foliations of $\mathfrak{M}$ with $(n-\ell)$-dimensional leaves. Since the $X_{\overline{\alpha}}$ are tangent to the fibers of π_1, and the X_α to those of π_2, the commutators

(1.10) $$L \sim ([X_\alpha, X_{\overline{\beta}}])$$

measure the extent to which these two foliations fail to fit together. A more useful consideration is the following. If we solve $r(z,\overline{w}) = 0$ for z^j, $1 \leqslant j \leqslant \ell$, in terms of z^α, $\ell+1 \leqslant \alpha \leqslant n$, and $\overline{w}$, then T_zQ_w is described by the slopes $p^j_\alpha = \partial z^j/\partial z^\alpha$. Hence, $(X_{\overline{\beta}}\, p^j_\alpha)$ measures how much T_zQ_w varies as w varies in Q_z. For $\ell = 1$, $r_{\overline{n}} \neq 0$, one obtains at $z = w \in M$

$$X_{\overline{\beta}}\, p_\alpha = \partial_{\overline{\beta}}(-r_\alpha/r_n) - (r_{\overline{\beta}}/r_{\overline{n}})\, \partial_{\overline{n}}(-r_\alpha/r_n). \tag{1.10a}$$

a well-known expression for the Levi form. For general codimension ℓ we consider the map

$$g:\ \mathfrak{M} \longrightarrow \mathbb{C}^n \times Gr(n-\ell, n), \qquad g(z, w) = (z, T_zQ_w). \tag{1.11}$$

It is not hard to see [12] that $dg \cong (dz, L)$, i.e. $(X_{\overline{\beta}}\, p^j_\alpha) \sim L$, so that g is an immersion precisely when M has non-degenerate Levi form. The image $g(\mathfrak{M})$ is then a smooth complex $(2n-\ell)$-submanifold, which inherits a reflection ρ from $(z, \overline{w}) \longrightarrow (w, \overline{z})$ on $\mathfrak{M}$. The fixed point set is precisely the set of planes $H_z(M)$, $z \in M$. If $\ell = 1$, $g(\mathfrak{M}) = \mathbb{C}^n \times \mathbf{P}^*_{n-1} \equiv P$ by reason of dimension. The bundle π: $P \longrightarrow \mathbb{C}^n$ may be identified with the bundle of lines in the cotangent space of $\mathbb{C}^n$ at each point.

2. Non-Degenerate Real Hypersurfaces

This theory has been studied extensively since the work of Chern and Moser [5]. We shall consider it here briefly from the point of view of complexification. We assume that $r_n \neq 0$, so that the single equation $r(z, \overline{w}) = 0$, for w fixed, defines z^n as a function of z^α (Greek indices run from 1 to n–1; the summation convention will be used). Differentiation with respect to z^α and then z^β gives

$$\begin{aligned} &1)\ r(z, \overline{w}) = 0, \\ &2)\ r_\alpha + p_\alpha r_n = 0, \\ &3)\ r_{\alpha\beta} + r_{\alpha n}\, p_\beta + r_{\beta n}\, p_\alpha + r_{nn}\, p_\alpha p_\beta + r_n\, \partial_\beta p_\alpha = 0. \end{aligned} \tag{2.1}$$

When the Levi form (1.10a) is non-degenerate, we can solve 1)-2) for $\overline{w}$ and substitute into 3) to obtain a holomorphic 2nd-order system of partial differential equations in the variables z. This is best written as a differential system on the bundle π: $P \longrightarrow \mathbb{C}^n$, which has p_α as fiber coordinates. We set

$$\theta = dz^n - p_\alpha dz^\alpha, \qquad \theta^\alpha = dz^\alpha, \qquad \theta_\alpha = dp_\alpha - F_{\alpha\beta} dz^\beta, \tag{2.2}$$

where $F_{\alpha\beta}$ is defined by (2.1) 3). The one form θ is determined up to a non-zero factor and vanishes on the fibers of π as well as on all $\widetilde{S} = \{T_zS;\ z \in S\}$, where $S \subset \mathbb{C}^n$ is a smooth complex hypersurface. The two systems (differential ideals),

$$I = \{\theta, \theta^\alpha\}, \qquad II = \{\theta, \theta_\alpha\}, \tag{2.3}$$

are both closed: $dI \subset I$, $dII \subset II$. The maximal integral varieties of $II = 0$ are the $\widetilde{Q}_z$. They are the images of the varieties $(Q_w, \overline{w})$ under g: $\mathfrak{M} \longrightarrow P$. The integral varieties of $I=0$ are the fibers of π, which are the images of the (z, Q_z) under g. The two systems I, II are interchanged by the reflection ρ, which multiplies θ by a non-zero factor. It is natural to pass to the complex line bundle $E \longrightarrow P$ of all non-zero multiples of θ. The main theorem is the following.

Theorem. *(Tresse-Cartan-Hachtroudi). Let Q_z be a holomorphic* n*-parameter family of analytic hypersurfaces in $\mathbb{C}^n$, such that through each point and tangent to each hyperplane there passes a unique Q_z. Then there exists an intrinsic subbundle* Y *of the coframe bundle of* E *with structure group covered by a subgroup* H *of* $SL(n+1, \mathbb{C})$. Y *carries an invariant normalized* $\mathfrak{sl}(n+1, \mathbb{C})$*-valued Cartan connection* π.

The analogous theorem for a smooth real family of hypersurfaces in $\mathbb{R}^n$ also holds. This is the context in which the three named authors worked. Tresse [11] considered the case $n=2$ and applied Sophus Lie's theory of invariants. His work was recast into the theory of projective connections and greatly simplified by E. Cartan [2]. In the book [7] Hachtroudi considered the case $n>2$ via Cartan projective connections. The above formulation follows that given in [4], [5].

The invariant theory of the real hypersurface M may be recovered by means of the reflection ρ. As noted, above, it acts on P with fixed point set $\widetilde{M} \cong M$, the locus of complex tangents $H_z(M)$. It induces a reflection on the complex line bundle E, with fixed point set a real line bundle $E_r \subset E$ over $\widetilde{M}$. As indicated above, this in turn induces a reflection on the bundle Y. The fixed point set is a real bundle $Y_r \subset Y$ over E_r with structure group covered by $H_r = H \cap S\mathcal{U}(p,q)$, where $(p-1, q-1)$ is the signature of the Levi form of M. When restricted to Y_r, the connection form π becomes the Chern-Moser connection. For further details see Faran [6], who actuallly works on the manifold $\mathfrak{M}$ rather than P. We shall be content here to consider the flat case, and then to indicate how the Moser normal form may be derived.

When the family Q_z is the set of (n–1)-planes in $\mathbb{C}^n$, we may view P as the set of "pointed" (n–1)-planes $\mathbf{P}_{n-1} \subset \mathbf{P}_n$. The geometry is best studied via projective frames

$$Z = (Z_0, Z_1, \ldots, Z_n) \in (\mathbb{C}^{n+1})^{n+1}, \qquad \det Z = 1. \tag{2.4}$$

Here Z_0 represents the base point of the hyperplane V, which is spanned by $Z_0, \ldots, Z_{n-1}$, and Z_n is a point off V. The subgroup H of the theorem is defined by the change of such adapted frames:

$$Z' = \mathcal{U}Z, \qquad \det \mathcal{U} = 1, \qquad \mathcal{U} \in H,$$

$$\begin{aligned} Z'_0 &= \mathcal{U}^0_0 Z_0 \\ Z'_\alpha &= \mathcal{U}^0_\alpha Z_0 + \mathcal{U}^\beta_\alpha Z_\beta \\ Z'_n &= \mathcal{U}^0_n Z_0 + \mathcal{U}^\beta_n Z_\beta + \mathcal{U}^n_n Z_n. \end{aligned} \tag{2.5}$$

The bundle Y is $SL(n+1)/K \longrightarrow P$ with structure group H/K, where K is the group of multiples of the identity matrix by the (n+1)-st roots of unity. The Maurer-Cartan forms on SL(n+1), given by

$$(2.6) \qquad dZ_j = \pi_j^i z_i, \qquad \mathrm{tr}\,\pi = 0, \qquad d\pi = \pi \wedge \pi,$$

yield the connection.

A *chain* is a variable-pointed hyperplane V_t: $t \in \mathbb{C}$ which moves in the simplest possible manner. Namely, the base point $Z_0(t) = Z_0 + tZ_n$ moves along the line ℓ spanned by Z_0, Z_n, and the plane V_t rotates about the skew-space spanned by Z_α, $1 \leqslant \alpha \leqslant n-1$. A parallel transport of the projective frame Z_α on V_t is defined by $dZ_\alpha = 0$. A projective parameter is defined by the cross ratio

$$(Z_0(t), A, B, C),$$

where A, B, C are independent points on ℓ.

We may add a non-degenerate hermitian form h on $\mathbb{C}^{n+1}$,

$$h(Z, \overline{Z}) = h_{i\overline{j}} \zeta^i \overline{\zeta}^j, \qquad Z = \zeta^i Z_i,$$

and let $S = \{Z \in \mathbf{P}_n\colon\ h(Z, \overline{Z}) = 0\}$. S is the (2n−1)-sphere if h has signature (n, −1) and is empty if h is positive definite. Regardless of the signature, $Q_Z = Z^\perp$ is the family of hyperplanes.

Let $S\mathcal{U}(h)$ be the set of frames for which $h_{i\overline{j}} = h(Z_i, \overline{Z}_j)$ has a canonical form, say, diagonal. It is invariant under the reflection given by the passage to the h-dual basis

$$(2.7) \qquad Z \longrightarrow W, \qquad h(Z_i, \overline{W}_j) = h_{i\overline{j}}.$$

On P this induces the map $Z_0 \in V \longrightarrow V^\perp \in Z_0^\perp$.

For a general (non-flat) structure $\{Q_z\}$, the connection π, which satisfies

$$(2.8) \qquad d\pi = \pi \wedge \pi + \Pi, \qquad \Pi = \text{curvature},$$

may be used to develop complex curves in Y into curves $t \longmapsto Z(t)$ in SL(n+1). Those for which the corresponding V_t moves along a chain project to the chains in P. Likewise, one can carry over the ideas of parallel translation and projective parameter to the general case.

These ideas permit us to define distinguished coordinate systems as follows. Let $t \longmapsto (z(t), p(t)) \in P$ be a chain with the (n−1)-plane $p(t)$ transverse to the curve $z(t)$. Let $Q(t)$ be the variety Q_z through $z(t)$ and tangent to $p(t)$. We choose t to be one of the distinguished projective parameters. It will then serve as the new z^n-coordinate. To get the new coordinates $z^1, \ldots, z^{n-1}$, we choose a parallel projective frame $Z_\alpha(t)$ on $Q(t)$ at $z(t)$. As observed by Faran [6] the curvature Π vanishes when restricted to each $\widetilde{Q}_z$, so that π induces a flat projective structure on each Q_z. This means that there are distinguished coordinates on Q_z determined up to fractional linear transformation.

On $Q(t)$ these coordinates are fixed by the choice of parallel frame $Z_\alpha(t)$ and give $z^1, \ldots, z^{n-1}$.

If initially $z(0) \in M$ and $p(0) = H_{z(0)}(M)$, then the uniqueness of chains, etc., with given initial data, and the reflection invariance of the equations imply that for t on a suitable real curve (to be transformed to the x^n-axis) $z(t) \in M$ and $p(t) = H_{z(t)}(M)$. The new coordinates yield a normal form for M about $z(0)$. This normal form may differ from Moser's in the third trace condition [5].

3. n-Manifolds in $\mathbb{C}^n$

We next consider an analytic real n-dimensional submanifold M of $\mathbb{C}^n$, which may be given locally as

$$M: r = (r^1, \ldots, r^n) = 0, \qquad \begin{array}{l} dr^1 \wedge \ldots \wedge dr^n \neq 0, \\ \partial r^1 \wedge \ldots \wedge \partial r^n = B\, dz_1 \wedge \ldots \wedge dz_n. \end{array} \tag{3.1}$$

We denote by N the set $r = 0$, $B = 0$, of points z at which M has a non-zero complex tangent space H_z. For $z \in N$ we assume

a) $\dim_{\mathbb{C}} H_z = 1$: H_z spanned by $X = \Sigma\, \xi^i\, \partial/\partial z_i$;

b) H_z is *non-degenerate*: either $XB = \Sigma\, \xi^i\, \partial_i B \neq 0$, or $\overline{X}B \neq 0$.

Then there is an invariant γ, first considered by Bishop [1],

$$\gamma = \frac{1}{2}\,|XB/\overline{X}B|, \qquad \gamma \in [0, \infty]. \tag{3.2}$$

The complex tangent is elliptic if $\gamma < \frac{1}{2}$, parabolic if $\gamma = \frac{1}{2}$, or hyperbolic if $\gamma > \frac{1}{2}$. If M is represented locally as a graph over its real tangent plane T_z

$$\begin{aligned} z_n &= F(z_1, x_2, \ldots, x_{n-1}) = az_1^2 + bz_1\overline{z}_1 + c\overline{z}_1^2 + x \cdot 0(1) + 0(2), \\ y_\alpha &= f_\alpha(z_1, x_2, \ldots, x_{n-1}) = 0(2), \qquad 2 \leqslant \alpha \leqslant n-1, \end{aligned} \tag{3.3}$$

then $\gamma = \gamma(0) = |c/b|$.

We seek holomorphic coordinates z in which the representation (3.3) has a particularly simple form. A principal result of the theory is the following.

Theorem. [9] *Assume* M *is real analytic and has an elliptic complex tangent at* z_0 *with* $\gamma(z_0) \neq 0$. *Then there exists a biholomorphic transformation taking* M *into the (implicit) form*

$$\begin{aligned} x_n &= z_1\overline{z}_1 + \Gamma(x_2, \ldots, x_n)(z_1^2 + \overline{z}_1^2), \qquad \Gamma(0) = \gamma, \\ y_2 &= \ldots = y_n = 0. \end{aligned} \tag{3.4}$$

In order to simplify the discussion of this result and other aspects of the theory, we shall restrict to the case $n = 2$. Then, provided $\gamma \neq \infty$, we may write M as

$$
(3.5) \qquad \begin{aligned} z_2 &= (q+H)(z_1, \overline{z}_1), & q &= z_1\overline{z}_1 + \gamma(z_1^2 + \overline{z}_1^2), \\ \overline{z}_2 &= (q+\overline{H})(\overline{z}_1, z_1), & H &= h + ik = O(|z_1|^3). \end{aligned}
$$

The osculating paraboloid, $x_2 = q$, $y_2 = 0$, is an elliptic paraboloid if $\gamma < \frac{1}{2}$, and contains the boundaries of the one-parameter family of analytic discs $x_2 = \text{const} \geqslant q(z_1)$, $y_2 = 0$.. Since these discs shrink to a point as the constant tends to zero, the paraboloid has the three-dimensional hull of holomorphy $x_2 \geqslant q$, $y_2 = 0$. A similar result holds if $\gamma < \frac{1}{2}$ and $H = \overline{H}$. This leads us to the biholomorphic flattening problem: find the new coordinates $z' = g(z)$ in which M lies in the real hyperplane $y_2' = \operatorname{Im} g_2(z) = 0$. Before considering this problem, we mention some of the negative results of [9]:

1) M: $z_2 = z_1\overline{z}_1 + z_1^2 + \overline{z}_1^2 + z_1\overline{z}_1(z_1 - \overline{z}_1)$ cannot be holomorphically flattened to third order ($\gamma = 1$).

2) There is a sequence $\{\gamma_j\}_{j=1}^{\infty}$, $\gamma_j > \frac{1}{2}$, dense in the interval $(\frac{1}{2}, \infty)$ of *exceptional values* of γ. For each γ_j there exists an M_j which can be flattened to j-th order, but not to (j+1)-th order.

3) M: $z_2 = z_1\overline{z}_1 + \gamma\overline{z}_1^2 + \gamma z_1^3\overline{z}_1$, $\frac{1}{2} < \gamma < \infty$, cannot be transformed into a real hyperplane by any (convergent) biholomorphic map. If $\gamma \neq \gamma_j$, then M can be formally flattened..

To flatten M, we must find a holomorphic function $g_2(z)$ with $g_2(z) = \overline{g}_2(\overline{z})$ on M. In other words, $f = g_2|_M = \overline{g}_2|_M$ is the restriction of both a holomorphic and an anti-holomorphic function. Hence, we must begin by characterizing the trace f on M of a function holomorphic on a neighborhood of M. This is best accomplished by complexification, $(z, \overline{z}) \longrightarrow (z, w)$,

$$
(3.6) \qquad \mathfrak{M}: \begin{aligned} z_2 &= (q+H)(z_1, w_1), \\ w_2 &= (q+\overline{H})(w_1, z_1). \end{aligned}
$$

The two projections (1.4) (replace η by w), when restricted to $\mathfrak{M}$ have the form

$$
(3.7) \qquad \begin{aligned} \pi_1(z_1, w_1) &= (z_1, (q+H)(z_1, w_1)), \\ \pi_2(z_1, w_1) &= (w_1, (q+\overline{H})(w_1, z_1)). \end{aligned}
$$

A basic fact, which we shall verify shortly by considering quadrics, is that the mappings (3.7) are (locally) two-fold branched coverings when $0 < \gamma \leqslant \infty$. The corresponding covering transformations, τ_2, τ_1, are holomorphic involutions on $\mathfrak{M}$, $\tau_1^2 = \tau_2^2 = \text{id}$. If $\tilde{f}$ denotes the analytic continuation of f to $\mathfrak{M}$, then

1) f holomorphic $\longleftrightarrow \tilde{f} = g \circ \pi_1 \longleftrightarrow \tilde{f} \circ \tau_2 = \tilde{f}$,

2) f anti-holomorphic $\longleftrightarrow \tilde{f} = g \circ \pi_2 \longleftrightarrow \tilde{f} \circ \tau_1 = \tilde{f}$.

Hence, the map τ_2 is a discrete analogue of the tangential Cauchy-Riemann operators (1.8).

So, to flatten M, we need to find a non-trivial function f on M satisfying $f \circ \tau_2 = f \circ \tau_1 = f$. The difficulty here is that τ_1 and τ_2 do not commute:

(3.8) $$\tau_1 \tau_2 \tau_1^{-1} \tau_2^{-1} = \varphi^2, \qquad \varphi = \tau_1 \tau_2.$$

φ^2 is a discrete analogue of the Levi form (1.10). Our approach to the flattening problem is to find new coordinates (ξ, η) on $\mathfrak{M}$ in which τ_1, τ_2 both have simple forms, and then to read off their invariants.

We first consider the case of the quadric (H = 0 in (3.5)), which will lead to a pair of linear involutions if $\gamma \neq 0$. (If $\gamma = 0$, $\pi_1(z_1, w_1) = (z_1, z_1 w_1)$ is birational and collapses the w_1-axis to a point.) We have $\tau_1(z, w) = (z', w')$, where $w' = w$, so

$$q(z_1, w_1) = w_2 = w_2' = q(z_1', w_1).$$

The solution $z_1' \neq z_1$ is $z_1' = -z_1 - \gamma^{-1} w_1$. This and a similar argument for π_2 gives, in matrix form,

$$\tau_1 = \begin{bmatrix} -1, & -\gamma^{-1} \\ 0, & 1 \end{bmatrix}, \qquad \tau_2 = \begin{bmatrix} 1, & 0 \\ -\gamma^{-1}, & -1 \end{bmatrix}, \qquad \varphi = \begin{bmatrix} -1+\gamma^{-2}, & \gamma^{-1} \\ -\gamma^{-1} & -1 \end{bmatrix}.$$

The characteristic equation for φ,

(3.10) $$\mu^2 + (2-\gamma^{-2})\mu + 1 = 0,$$

with discriminant $\gamma^{-4}(1-4\gamma^2)$, has the roots $\mu, \bar{\mu}, \mu^{-1}$. In the elliptic case $\mu = \bar{\mu}$, and the mapping φ is hyperbolic, having eigenvalues μ and μ^{-1}. In the hyperbolic case $\mu \neq \bar{\mu} = \mu^{-1}$, so that $|\mu| = 1$ and φ is an elliptic mapping. Degeneracy in the Levi form corresponds to $\varphi^2 = \mathrm{id}$, which happens when $\gamma = \infty$, or more generally to φ being nilpotent, which happens precisely when μ is a root of unity. Solving (3.10) for γ in terms of μ shows that the corresponding γ's, the exceptional γ's, are dense in $(\frac{1}{2}, \infty)$.

The basic result on pairs of linear involutions is the following.

Lemma 1) [9]. *If the linear involutions τ_1, τ_2 on $\mathbb{C}^2$ have no common eigenvectors, then there exist linear coordinates (ξ, η) in which*

$$\tau_1(\xi, \eta) = (\lambda\eta, \lambda^{-1}\xi), \qquad \tau_2(\xi, \eta) = (\lambda^{-1}\eta, \lambda\xi).$$

τ_1 and τ_2 have a common eigenvector in the parabolic case.

Next, we consider a pair of non-linear analytic involutions τ_i, $i = 1, 2$, on $\mathbb{C}^2$, fixing the origin O with $d\tau_1(0)$, $d\tau_2(0)$ having no common eigenvectors. By Lemma 1 we may assume they have the form

$$(3.11) \qquad \tau_i: \begin{array}{ll} x' = \lambda_i y + f_i(x, y), & f_i, g_i = O(2), \\ y' = \lambda_i^{-1} x + g_i(x, y), & \lambda_2 = \lambda_1^{-1}. \end{array}$$

From $\tau_i^2 = \text{id}$ follows $f_i \circ \tau_i = -\lambda_i g_i$, so that, in essence, there are only two independent functions involved. We seek a (formal) coordinate change

$$(3.12) \qquad \psi: \begin{array}{l} x = \xi + p(\xi, \eta) \\ y = \eta + q(\xi, \eta), \end{array}$$

so that $\tau_i^0 = \psi^{-1} \tau_i \psi$ has a simple (normal) form. ψ is said to be normalized if q has no terms of type $\xi^i \eta^{i+1}$ and p has no terms of type $\xi^{i+1} \eta^i$.

Lemma 2) [9]. *Let* τ_i, $i = 1, 2$, *be an analytic pair of (local) involutions on* $\mathbb{C}^2$ *given by* (3.11) *with* $\lambda_1 \lambda_2^{-1}$ *not a root of unity. Then there exists a unique normalized formal transformation* ψ *as in* (3.12), *such that the* $\tau_i^0 = \psi^{-1} \tau_i \psi$ *have the form*

$$\tau_i^0(\xi, \eta) = (\Lambda_i \eta, \Lambda_i^{-1} \xi),$$

where $\Lambda_i = \Lambda_i(\xi\eta) = \lambda_i + \ldots$. *If* $|\lambda_1| \neq |\lambda_2|$, *then* ψ *and the factors* Λ_i *converge.*

For $\lambda_2^{-1} = \lambda_1 \equiv \lambda$, $\mu = \lambda^2$, and we have convergence when $|\mu| \neq 1$, which corresponds to $0 < \gamma < \frac{1}{2}$ via (3.10).

In order to make Lema 2) applicable to the surface M, one must also take into account the anti-holomorphic involution $\rho(z, w) = (\overline{w}, \overline{z})$, which carries $\mathfrak{M}$ into itself and fixes M pointwise. It is easy to see that $\rho \tau_1 = \tau_2 \rho$. It turns out [9] that with ρ suitably normalized, one can make $\Lambda_2^{-1} = \Lambda_1 \equiv \Lambda$, and $\Lambda = \overline{\Lambda}$ if $\gamma < \frac{1}{2}$, or $\Lambda \overline{\Lambda} = 1$ if $\gamma > \frac{1}{2}$ and not exceptional. One then introduces new coordinates by

$$\begin{aligned} z_1 &= i\Lambda^{-1/2}(\Lambda \xi + \eta), \\ w_1 &= -i\Lambda^{-1/2}(\xi + \Lambda \eta), \qquad (3.12) \\ z_2 &= w_2 = (\Lambda - \Lambda^{-1})^2 (\Lambda + \Lambda^{-1})^{-1} \xi \eta. \end{aligned}$$

In the coordinates (z_1, z_2) on $\mathbb{C}^2$ one has the form (3.4) $(n = 2)$ for M. If one drops the requirement that ψ be normalized, then the additionl freedom allows one to achieve [9], for $n = 2$,

$$(3.13) \qquad \Gamma(x_2) = \gamma + \delta x_2^s,$$

where $\delta = \pm 1$, 0, and $s \in \mathbb{Z}^+$.

Thus, in the case $0 < \gamma < \frac{1}{2}$, the biholomorphic equivalence problem is completely solved, (γ, δ, s) being a complete system of invariants. Note that the normal form (3.4), (3.13) is algebraic. This is definitely not the case for real hypersurfaces [5]. Although we have (discrete) analogues here of the tangential Cauchy-Riemann operators and of the Levi form, there seem to be no analogues of the frame bundle Y or of the connection π. In fact, the Cartan method of equivalence apparently is not applicable to this problem.

REFERENCES

[1]. E. Bishop, *Differentiable Manifolds in Complex Euclidean Space,* **Duke Math. Jour.** 32 (1965), 1-22.

[2]. E. Cartan, *Sur les varietés à connexion projective,* **Bull. Soc. Math. France** 32 (1924), 205-241.

[3]. E. Cartan, *Sur la géométrie pseudo-conforme des hypersurfaces de l'espace de deux varables complex I,* **Ann. Mat. Pure Appl.** 11 (1932), 17-90, and **II Ann. Scuola Norm. Sup. Pisa, Sci. Fis. Mat.** 1 (1932), 333-354.

[4]. S. S. Chern, *On the Projective Structure of a Real Hypersurface in* $\mathbb{C}^n$, **Math. Scand.** 36 (1975), 74-82.

[5]. S. S. Chern and J. K. Moser, *Real Hypersurfaces in Complex Manifolds,* **Acta Math.** 133 (1974), 219-271.

[6]. J. J. Faran, *Segre Families and Real Hypersurfaces,* **Invent. Math.** 60 (1980), 135-172.

[7]. M. Hachtroudi, *Les espaces d'éléments a connexion projective normale,* **Act. Sci. et Ind.** No. 565, Hermann, Paris, 1937.

[8]. G. Harris, *The Traces of Holomorphic Functions on Real Submanifolds,* **Trans. AMS** 242 (1978), 205-223.

[9]. J. K. Moser and S. M. Webster, *Normal Forms for Real Surfaces in* $\mathbb{C}^2$ *near Complex Tangents and Hyperbolic Surface Transformations,* **Acta Math.** (To appear).

[10]. B. Segre, *Intorno al problem di Poincaré della rapprezentazione pseudo-conform,* **Rend. Acc. Lincei** 131 (1931), 676-683.

[11]. A. Tresse, *Détermination des invariant ponctuels de l'equation differenttielle ordinaire du second ordre* $y'' = \omega(x, y, y')$, Memoire couronné par l'**Academie Jablonowski,** S. Hirkel, Leipzig, 1896.

[12]. S. Webster, *Holomorphic Mappings of Domains with Generic Corners,* **Proc. AMS** 86 (1982), 236-240.

School of Mathematics
University of Minnesota
Minneapolis, Minnesota 55455
U.S.A.

A BOGGESS*

A survey of recent CR extension results

In this article we survey some of the recent results on CR extension and report on some current results obtained in [BP]. This article is an expanded version of a talk the author gave at the Workshop in Several Complex Variables in August of 1983, sponsored by the Department of Mathematics, Centro de Investigacion y Estudios Avanzados del IPN, Mexico City.

Let us fix notation. Let M be a real CR submanifold of $\mathbb{C}^n$ of class at least C^2 with $\operatorname{codim}_{\mathbb{R}} M = d$. We let $T(M)$ (resp. $T^{\mathbb{C}}(M)$) be the tangent bundle (resp. complexified tangent bundle) to M. Let $H(M) \subset T(M)$ (resp. $H^{\mathbb{C}}(M)$) be the holomorphic (resp. complexified holomorphic) tangent bundle to M. We will always work in the generic case, which means that for p in M, $\dim_{\mathbb{R}} H_p(M)$ is minimal (i.e., $\dim_{\mathbb{R}} H_p(M) = 2n-2d$). The quotient $T_p(M)/H_p(M)$ will be identified with the totally-real tangent space of M at p and will be denoted by $Y_p(M)$. The complex structure map J on $T_p(\mathbb{C}^n)$ can be restricted to $Y_p(M)$ to yield an isometry between $Y_p(M)$ and $N_p(M)$: = the space of vectors in $\mathbb{C}^n$ which are orthogonal to $T_p(M)$ under the usual Euclidean metric on $\mathbb{R}^{2n} = \mathbb{C}^n$. The map J: $H_p^{\mathbb{C}}(M) \longrightarrow H_p^{\mathbb{C}}(M)$ has eigenspaces $H_p^{1,0}(M)$ and $H_p^{0,1}(M)$ corresponding to the eigenvalues i and $-i$, respectively.

The first Levi form is a map L_p^1: $H_p^{1,0}(M) \longrightarrow Y_p(M)$ and is defined as follows. If $X_p \in H_p^{1,0}(M)$, then

$$L_p^1(X_p) := \frac{1}{2i}\pi_p[X, \overline{X}]_p$$

where $X \in H^{1,0}(M)$ is any vector field extension of X_p and π_p: $T_p^{\mathbb{C}}(M) \longrightarrow Y_p^{\mathbb{C}}(M)$ is the projection map. Since $\overline{L_p^1(X_p)} = L_p^1(X_p)$, clearly L_p^1 maps into $Y_p \otimes_{\mathbb{R}} 1$, which we identify with Y_p.

It will be useful to have an extrinsic version of the Levi form $\widetilde{L}_p^1$: $H_p^{1,0}(M) \longrightarrow N_p(M)$ which is defined as $\widetilde{L}_p^1 = J \circ L_p^1$. Suppose $p = 0$ and M is given near 0 as $M = \{(z, w) \in \mathbb{C}^{d+m};\ \operatorname{Re} z = h(\operatorname{Im} z, w)\}$ where h: $\mathbb{R}^d \times \mathbb{C}^m \longrightarrow \mathbb{R}^d$ is a C^2-function. It can be shown that if

$$X = \sum_{j=1}^{m} a_j \frac{\partial}{\partial w_j} \in H_0^{1,0}(M), \quad a_j \in \mathbb{C},$$

then

$$\widetilde{L}_0^1(X_0) = \sum_{j,k=1}^{m} \frac{\partial^2 h}{\partial w_j \partial \overline{w}_k}(0)\, a_j \overline{a}_k \in \mathbb{R}^d. \tag{1}$$

Here we are identifying $N_0(M)$ with a copy of $\mathbb{R}^n$.

Let $\Gamma_p(M)$ be the convex hull in $N_p(M)$ of the image of $\widetilde{L}_p^1$. From (1), it is clear that $\Gamma_p(M)$ is a

*Research supported by NSF Grant MSC-8301369.

convex cone in $N_p(M)$. Since $\widetilde{L}^1_p$ depends on the second derivatives of a local defining function for M, Γ_p represents the set of directions of "bending or twisting" in the following sense; if Γ_p is contained in (real) subspace A_q of $N_p(M)$ with $\dim_{\mathbb{R}} A_q = q$, $0 \leqslant q \leqslant d$, then modulo terms which vanish at p to third order, M can be contained in the linear subspace $T_p(M) \oplus A_q$ which has real dimension $\dim_{\mathbb{R}} M + q$. If the interior of Γ_p (relative to $N_p(M)$) is nonempty, then M cannot be contained in any smaller subspace of $\mathbb{C}^n$ (even if one mods out third-order terms). In this case it is a theorem that CR functions on M locally extend to holomorphic functions on an open subset of $\mathbb{C}^n$. For the precise statement of the theorem, we need one more piece of terminology. If Γ_1 and Γ_2 are subcones of $N_p(M)$, we say that $\Gamma_1 < \Gamma_2$ if $\Gamma_1 \cap S \subset\subset \Gamma_2 \cap S$ where S is the unit sphere in $N_p(M)$ centered at p.

Theorem (cf. Theorem 1.1 in [BPo]). *Suppose* M *is a* CR, *generic submanifold of* $\mathbb{C}^n$ *of class* C^3. *Let* $p \in M$ *such that* Γ_p *has nonempty interior relative to* $N_p(M)$. *Given* $\omega \ni p$ *an open subset of* M, *there exists an open set* $\omega' \subset M$ *and an open set* Ω *of* $\mathbb{C}^n$ *with the following properties.*

i) $p \in \omega' \subset \overline{\Omega} \cap M \subset \omega$.

ii) Given a cone $\Gamma < \Gamma_p$, *there is an open set* $\omega_1 \ni 0$ *in* M *and an* $\epsilon > 0$ *such that* $\omega_1 \oplus \{\Gamma \cap B(0,\epsilon)\} \subset \Omega$ *where* $B(0,\epsilon)$ *is the ball of radius* ϵ *centered at the origin.*

iii) If f *is a continuous* CR *function on* ω, *then there is a continuous function* F *on* $\overline{\Omega}$ *which is holomorphic on* Ω *with* $F = f$ *on* ω'.

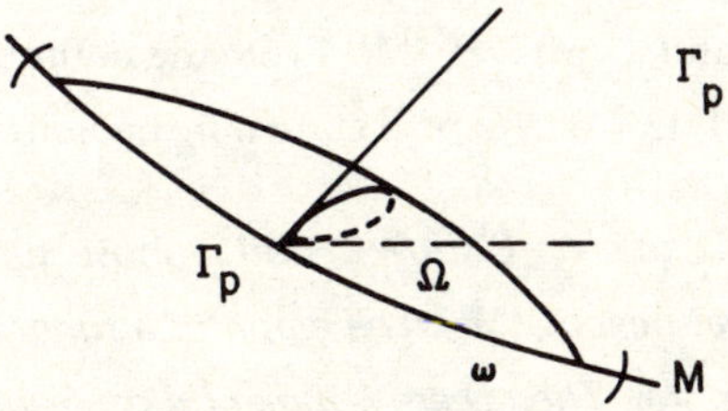

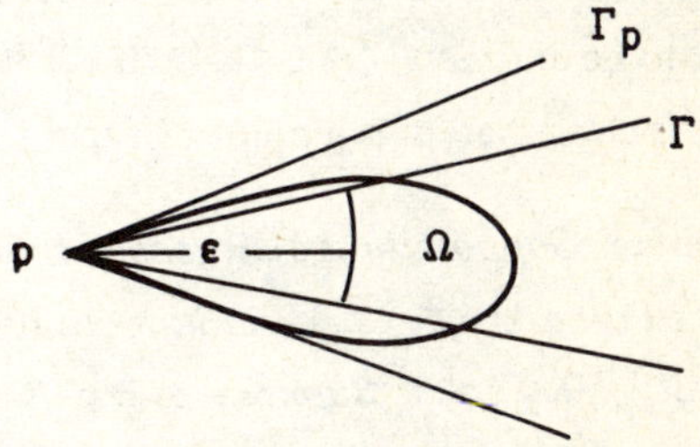

The theorem essentially states that CR functions on M near p extend to holomorphic functions on an open set Ω and Γ_p governs the size of Ω. In particular, Part ii) yields that Ω locally contains any cone which is smaller than Γ_p. If $\Gamma_p = N_p(M)$, then $\Gamma_p < \Gamma_p$ and so Ω contains p in its interior. This result carries with it a regularity result for CR functions; namely, that an apriori continuous CR function must be as smooth as M.

Earlier CR extension results such as [Gr] and [HW] obtained a corresponding Ω, but they had no information on the size of Ω. Theorem I has also been recently obtained by [BCT] using completely different techniques. To obtain their CR Extension Theorem, one must combine Theorems 6.1 and 2.3 (Chapter II) in their paper.

In the case when the interior of Γ_p relative to $N_p(M)$ is empty (i.e., Γ_p is contained in a lower dimensional subspace of $N_p(M)$), CR functions generally do not extend as holomorphic functions. For example, if M is contained in a Levi flat hypersurface N, then $\dim \Gamma_p \leq \operatorname{codim}_{\mathbb{R}} M - 1$ and the hull of holomorphy of any subset of M is contained in N. However, CR functions on M near p may extend to CR functions on a larger submanifold $\widetilde{M}$ with boundary. This situation is well understood for the case when Γ_p contains a one-dimensional ray and $\dim_{\mathbb{R}} \widetilde{M} = \dim_{\mathbb{R}} M + 1$. In [HT, Theorem 9.1], it was shown that if $\nu \in N_p(M)$ is in the image of $\widetilde{L}^1_p$, then CR functions on M near p extend to CR functions on a submanifold $\widetilde{M}$ with boundary. In [HT], $\widetilde{M}$ extends in the direction ν and is roughly one third as smooth as M. In [AB], this result was improved to allow $\nu \in \Gamma^1_p$ and the corresponding $\widetilde{M}$ was shown to be roughly half as smooth as M. Now in [BP], this result was further improved and the corresponding $\widetilde{M}$ was shown to be almost as smooth as M (see the precise statement of our theorem below). Moreover, in [BP], the authors consider higher-order Levi forms which we will soon discuss. In [BP], it is only required that one of the Levi forms of order at least one be nonvanishing.

To discuss the situation in [BP], let us first define the type of a point. Assume that M is of class C^k, $k \geq 2$. For $p \in M$, let $\mathcal{L}^0_p(M) := H^{\mathbb{C}}_p(M)$ and for $1 \leq j \leq k-1$, let $\mathcal{L}^j_p(M)$ be the subspace of $T^{\mathbb{C}}_p(M)$ generated by $H^{\mathbb{C}}_p(M)$ and all Lie brackets at p of order at most j of elements in $H^{\mathbb{C}}(M)$. We have $H^{\mathbb{C}}_p(M) = \mathcal{L}^0_p(M) \subset \mathcal{L}^1_p(M) \subset \ldots \subset \mathcal{L}^{k-1}_p(M) \subset T^{\mathbb{C}}_p(M)$. We do *not* assume that $\dim_{\mathbb{R}} \mathcal{L}^j_p(M)$ is constant in p. Indeed, most of the interesting examples occur when $\dim_{\mathbb{R}} \mathcal{L}^j_p(M)$ is nonconstant in p. A point $p \in M$ is said to be of *type* ℓ, $1 \leq \ell \leq k-1$, if $\mathcal{L}^{\ell-1}_p(M) = H^{\mathbb{C}}_p(M)$ and $\mathcal{L}^{\ell}_p(M) \supset H^{\mathbb{C}}_p(M)$. From the definition of L^1_p, it is clear that p is a point of type 1 if and only if $L^1_p \not\equiv 0$. We now state our theorem.

II **Theorem.** *Suppose* M *is a* CR *generic submanifold of* $\mathbb{C}^n$ *of class* C^k, $k \geq 3$, *and suppose* $p \in M$ *is a point of type* ℓ, $1 \leq \ell \leq k-1$. *Suppose furthermore that near* p, M *is the graph of a function of class* $C^{k,\epsilon}$, $0 < \epsilon \leq 1$. *Suppose* $\omega \ni p$ *is an open set in* M. *Then there is a manifold* $\widetilde{\omega}$ *with boundary of class* C^k *with the following two properties.*

i) $\operatorname{Dim}_{\mathbb{R}} \widetilde{\omega} = \dim_{\mathbb{R}} M + 1$ *and the boundary of* $\widetilde{\omega}$ *is an open subset of* M *which contains* p.

ii) Each CR *function of class* C^j, $0 \leq j \leq k-1$, *on* ω *has a* CR *extension of class* C^j *on* $\widetilde{\omega}$.

We can also give a more precise description of $\widetilde{\omega}$. To do this, we must define the Levi forms. For a point $p \in M$ of type at least ℓ, we define the *ℓth* Levi form L^{ℓ}_p as a map $L^{\ell}_p : H^{1,0}_p(M) \longrightarrow N_p(M)$. If $X_p \in H^{1,0}(M)$, then

$$L_p^\ell(X_p) := \frac{1}{2i}\pi_p \circ J\Big\{\sum_{\epsilon_1,\ldots,\epsilon_{\ell-1}} C_\epsilon [X^{\epsilon_1}, [X^{\epsilon_2}, \ldots, [X^{\epsilon_{\ell-1}}, [X, \overline{X}]\ldots]_p\Big\}$$

Here $X \in H^{1,0}(M)$ is a vector field extension of X_p; J is the complex structure map on $T_p^{\mathbb{C}}(\mathbb{C}^n)$, and π_p is the orthogonal projection map onto $N_p(M)$. The sum is taken over all $\epsilon = (\epsilon_1, \ldots, \epsilon_{\ell-1})$ with $\epsilon_1 = 1$ or -1 with the convention that $X^1 = X$ and $X^{-1} = \overline{X} \in H^{0,1}(M)$. Finally $C_\epsilon = 1/(q+1)!\,(\ell-q)!$, where q is the number of occurrences of 1 in the set $\{\epsilon_1, \ldots, \epsilon_{\ell-1}\}$. The choice of C_ϵ is for technical convenience only. If p is a point of type at least ℓ, then $L_p^\ell(X_p)$ will not depend on the choice of the vector field extension X. In [BP], we show that if p is a point of type at least ℓ, then p is a point of type ℓ if and only if $L_p^{\ell-1} \equiv 0$ and L_p^ℓ is not identically zero. Also in [BP], we give a computational formula for L_p^ℓ in terms of the $(\ell+1)st$-order complex Hessian of a set of local defining functions for M.

Namely, suppose $p = 0$ and $M = \{(z, w);\ \operatorname{Re} z = h(\operatorname{Im} z, w)\}$. If $p = 0$ is a point of type ℓ and $X_0 = \sum_{j=1}^m a_j \frac{\partial}{\partial w_j} \in H^{1,0}(M)$, then

(2)

$$L_0^\ell(X_0) = \sum_{q=0}^{\ell-1} \frac{1}{(q+1)!\,(\ell-q)!} \sum_{j,k=1}^m \sum_{\substack{|\alpha|=q \\ |\beta|=\ell-q-1}} \binom{\ell-1}{\alpha_1 \cdots \alpha_m\ \beta_1 \cdots \beta_m} \cdot \frac{\partial^{\ell+1} h(0)}{\partial w_1^{\alpha_1} \ldots \partial w_m^{\alpha_m} \partial w_j \partial \overline{w}_1^{\beta_1} \ldots \partial \overline{w}_m^{\beta_m} \partial \overline{w}_k} \cdot A_1^{\alpha_1} \ldots A_m^{\alpha_m} A_j \overline{A}_1^{\beta_1} \ldots \overline{A}_m^{\beta_m} \overline{A}_k$$

Let $\Gamma_p^{\widetilde{\omega}}$ be the line which lies tangent to $\widetilde{\omega}$ at P but normal to M. Suppose $\omega \supset B(p, r) \cap M$ where $B(p, r)$ is the open ball in $\mathbb{C}^n$ of radius r centered at p. If $\nu \in N_p(M)$ lies in the image of L_p^ℓ, then the proof of our theorem shows that $\widetilde{\omega}$ can be chosen so that the angle between ν and $\Gamma_p^{\widetilde{\omega}}$ converges to zero as $r \downarrow 0$. In other words, $\widetilde{\omega}$ extends roughly in the direction $\nu \in \operatorname{image}[L_p^\ell]$ Moreover, we show the diameter of $\widetilde{\omega} \cap M$ is proportional to r and that the diameter of $\widetilde{\omega}$ along $\Gamma_p^{\widetilde{\omega}}$ is proportional to $r^{\ell+1}$ (ℓ = type of p).

If M is a real hypersurface, then $N_p(M)$ is a copy of the real line. If $L_p^\ell \not\equiv 0$, then it is clear from (2) that $\operatorname{image}[L_p^\ell] = N_p(M)$ if ℓ is even and $\operatorname{image}[L_p^\ell]$ is at least a ray if ℓ is odd. Thus our theorem gives the following result for real hypersurfaces.

If $p \in M$ is a point of type ℓ, $1 \leq \ell \leq k-1$, then CR functions on M near p extend to holomorphic functions on an open set $\widetilde{\omega}$ in $\mathbb{C}^n$. If ℓ is even, then $\widetilde{\omega}$ contains p (i.e., $\widetilde{\omega}$ lies on both sides of M), and if ℓ is odd, then $\widetilde{\omega}$ lies (at least) to one side of M (the side given by $\operatorname{image}[L_p^\ell]$).

This result on hypersurfaces has also been announced by Baouendi and Treves [BT2], and independently by Rea.

The proof of our theorem uses the technique of analytic discs. An *analytic disc* is a continuous map $A\colon \overline{D} \longrightarrow \mathbb{C}^n$ ($D :=$ the unit disc in $\mathbb{C}$) which is holomorphic on D. The *boundary* of A is defined as the restriction of A to S^1 := the unit circle. We identify the analytic disc with its image in $\mathbb{C}^n$ and likewise for the boundary of A. We construct $\tilde{\omega}$ as a union of (pieces of) analytic discs with boundaries in ω. To find the extension of a given CR function on ω, we first uniformly approximate the CR function (on a slightly smaller ω) by means of a sequence of entire functions (cf. [BT1, Theorem 2.1]), and then we use the maximum principle on the discs to conclude that the approximating sequence of entire functions converges to a CR function on $\tilde{\omega}$.

Now we outline the construction of the analytic discs which fill out $\tilde{\omega}$. We suppose the given p is the origin (O) and that $\operatorname{codim}_{\mathbb{R}} M = d$. We can write $\mathbb{C}^n = \mathbb{C}^d \times \mathbb{C}^m$ with coordinates $(x+iy, w)$, $x+iy \in \mathbb{C}^d$, $w \in \mathbb{C}^m$. Locally, we describe M as $M = \{x = h(y, w)\}$ where $h: \mathbb{R}^d \times \mathbb{C}^m \longrightarrow \mathbb{R}^d$ is a function of class $C^{k,\epsilon}$ with $h(0) = 0$ and $Dh(0) = 0$.

We are hunting for analytic discs of the form $A = (G, W): \overline{D} \longrightarrow \mathbb{C}^d \times \mathbb{C}^m$ with boundary in M. Now boundary $(A) \subset M$ is equivalent to

$$\text{(3)} \qquad \operatorname{Re} G(\zeta) = h(\operatorname{Im} G(\zeta), W(\zeta)), \qquad |\zeta| = 1$$

From this equation we see that we are at liberty to choose W. Once W is chosen, G must be found so that (3) is satisfied. If we apply the Hilbert transform to both sides of (3), we eliminate Re G to obtain

$$v = T(H(v, W) + c \qquad \text{on } S^1.$$

where $v := \operatorname{Im} G|_{S^1}$ and $H(v, W)(\zeta) := h(v(\zeta), W(\zeta))$ and where $c = \operatorname{Im} G(\zeta = 0)$. The above equation is called Bishop's equation [B]. Conversely, if $W: \overline{D} \longrightarrow \mathbb{C}^m$ and $c \in \mathbb{R}^d$ are given and $v: S^1 \longrightarrow \mathbb{R}^d$ satisfies Bishop's equation, then the theory of the Hilbert transform guarantees that $H(v, W) + iv: S^1 \longrightarrow \mathbb{C}^d$ is the boundary values of an analytic disc $G: \overline{D} \longrightarrow \mathbb{C}^d$ with $\operatorname{Im} G(\zeta = 0) = c$. In particular, (3) is satisfied, and so if $A := (G, W): \overline{D} \longrightarrow \mathbb{C}^n$, then the boundary of A is contained in M. Thus, a solution to Bishop's equation generates an analytic disc with boundary in M.

It is convenient to take c as a parameter with the following format. Suppose the parameters $y \in \mathbb{R}^d$ and $z \in \overline{D}$ are given. We let $c = y - P(z, T(H(v, W)))$, where $P(z, u)$ is the Poisson integral of the function u evaluated at z. This leads to a modified Bishop's equation

$$\text{(9)} \qquad v = T(H(v, W)) + y - P(z, T(H(v, W))).$$

Note that the function $V(\zeta) = P(\zeta, T(H(v, W))) + y - P(z, T(H(v, W)))$, $\zeta \in \overline{D}$, is harmonic and $V|_{S^1} = v$. Thus

$$\text{(5)} \qquad V(\zeta = z) = y$$

This will be convenient in later computations.

The heart of the technical analysis is the construction of the solution to the modified Bishop's equation with smooth dependence on the parameters y, z, and the analytic disc W. We introduce more notation. Fix $0<\alpha<1$. We let $\mathcal{O}^{j,\alpha}(\overline{D})$ be the space of analytic discs $W\colon \overline{D} \longrightarrow \mathbb{C}^m$ such that $W|_{S^1} \in C^{j,\alpha}(S^1)$. This is a Banach space under the $C^{j,\alpha}(S^1)$ norm (completeness follows from the maximum principle). Our solution to the modified Bishop's equation is summarized in the following lemma (cf. [BP] for the proof).

LEMMA. *Suppose* h *is of class* $C^{k,\epsilon}$ *and fix* α, j, $0<\alpha<1$, $0\leq j\leq k$. *There exists a* $\delta = \delta(\alpha,\epsilon)>0$ *and there exist neighborhoods* $\mathcal{W}^{k,\alpha}\subset \mathcal{O}^{k,\alpha}(\overline{D})$ *and* $\mathcal{Y}\subset \mathbb{R}^d$ *(containing the origins) and a map* $v\colon \mathcal{W}^{k,\alpha}\times\mathcal{Y}\times\overline{D} \longrightarrow C^{j,\delta}(S^1)$ *which is of class* C^{k-j} *(uniformly) in the sense of Banach spaces such that for each* $W\in\mathcal{W}^{k,\alpha}$, $y\in\mathcal{Y}$, $z\in\overline{D}$, $v(W,y,z)(\cdot)$ *is the unique solution to equation* (4).

We now give a proof of theorem II in a special case. We write $\mathbb{C}^n = \mathbb{C}^d\times\mathbb{C}^m$ $(n=m+d)$ with coordinates (z,w) with $z=x+iy$ and assume $p=0$ and $M=\{(z,w); x=h(y,w)\}$. We shall further assume that h is independent of y and a homogeneous polynomial in w and $\overline{w}$ with no pure terms (a term is pure if it only involves w or $\overline{w}$). From the hypothesis on $\widetilde{L}^{\ell}_0$, there must exist a nonzero vector $\nu\in N_0(M)$ and a vector $A\in H^{1,0}_0(M)$ with $\widetilde{L}^{\ell}_0(A)=\nu$. Now in these coordinates $H^{1,0}(M)$ can be identified with the set $\{(0,w)\in\mathbb{C}^d\times\mathbb{C}^m; w=(w_1,\dots,w_m)\in\mathbb{C}^m\}$. By a complex linear change of coordinates in w only, we may assume $A=(1,0,\dots,0)$. From (2), we have

$$\sum_{q=0}^{\ell-1}\frac{1}{(q+1)!\,(\ell-q)!}\binom{\ell-1}{q}\frac{\partial^{\ell+1}h(0)}{\partial w_1^{q+1}\,\partial\overline{w}_1^{\ell-q}} = \nu. \tag{6}$$

Let us consider Bishop's equation with the following family of analytic discs for W

$$W(r,\lambda,w)(\zeta) = \begin{bmatrix} w_1+\dfrac{r(\lambda+\zeta)}{1+\lambda} \\ w_2 \\ \vdots \\ w_m \end{bmatrix}, \qquad |\zeta|\leq 1.$$

Here, r, λ, w are parameters with the following restrictions $r\geq 0$, $0\leq\lambda\leq 1$, $w\in\mathbb{C}^m$. Now the map

$$(r,\lambda,w)\longmapsto W(r,\lambda,w)(\cdot)\in C^{k,\alpha}(S^1)$$

is clearly C^∞. Moreover,

$$|W(r,\lambda,w)(\cdot)|_{k,\alpha}\leq r+|w|$$

where $|\cdot|_{k,\alpha}$ denotes the standard norm on the space $C^{k,\alpha}$.

Let $\qquad T_{r_0}=\{(r,\lambda,y,w);\ 0\leq\lambda\leq 1,\ r\leq r_0,\ \max|y|,|w|<r\}.$

Then $$|W(r, \lambda. w)(\cdot)|_{k,\alpha} \leqslant 2r_0 \qquad \text{on } T_{r_0}.$$

We restrict r_0 so that $W(r, \lambda, w)(\cdot) \in \mathcal{W}^{k,\alpha}$ and $y \in \mathcal{Y}$ for $(r, \lambda, y, w) \in T_{r_0}$, where $\mathcal{W}^{k,\alpha}$ and $\mathcal{Y}$ are the sets in the lemma. We obtain a solution v to Bishop's equation

$$v = T(H(W)) + y - P(z, T(H(W))). \tag{7}$$

The function v depends on the parameters $(r, \lambda, y, w) \in T_{r_0}$ and on the parameter $z \in \overline{D}$, and v takes values in $C^{j,\delta}(S^1)$ where $0 \leqslant j \leqslant k$ and $\delta > 0$. From the lemma, we see that

$$v: T_{r_0} \times \overline{D} \longmapsto C^{j,\delta}(S^1)$$

is a map of class C^{k-j} (uniformly) $0 \leqslant j \leqslant k$.

We need to have an estimate on $|v|_{0,\delta}$. First, note that

$$|v|_{0,\delta} \leqslant |T|_{0,\delta} |H(W)|_{0,\delta} + |y| + |P(z, T(H(W)))|$$

where $|T|_{0,\delta}$ refers to the operator norm of the map $T: C^{0,\delta} \longrightarrow C^{0,\delta}$. By the maximum principle for harmonic functions,

$$|P(z, T(H(W)))| \leqslant |T(H(W))|_{\infty} \leqslant |T|_{0,\delta} |H(W)|_{0,\delta}.$$

Therefore

$$|v|_{0,\delta} \leqslant 2|T|_{0,\delta} |H(W)|_{0,\delta} + |y|.$$

Since $C^{0,\delta}$ is a Banach algebra and $h(w)$ is a polynomial of degree $\ell+1$, there is a uniform constant C with

$$|v|_{0,\delta} \leqslant 2C |T|_{0,\delta} |W|_{0,\delta}^{\ell+1} + |y|.$$

Since $|y| < r_0$ and $|W(r, \lambda, w)(\cdot)|_{0,\delta} \leqslant 2r_0$ for $(r, \lambda, y, w) \in T_{r_0}$, clearly there is a constant $C = C(\delta)$ with

$$|v|_{0,\delta} \leqslant C r_0^{\ell+1} + r_0. \tag{8}$$

From our discussion of the theory of Bishop's equation, we see that the solution v to equation (7) generates an analytic disc $G = U + iV: \overline{D} \longmapsto \mathbb{C}^n$ with $V|_{S^1} = v$ and

$$U(\zeta) = H(W)(\zeta), \qquad |\zeta| = 1. \tag{9}$$

If we define the analytic disc $A := (G, W): \overline{D} \longmapsto \mathbb{C}^n$ then we have boundary $(A) \subset M$ by (9). Since W and v depend on the parameters r, λ, y, w, z, G (and hence A) also depend on the same parameters. Since the map $v\colon T_{r_0}\times\overline{D} \longmapsto C^{j,\delta}(S^1)$ is of class C^{k-j}, it is not hard to show that $A\colon T_{r_0}\times\overline{D} \longrightarrow C^{j,\delta'}(\overline{D})$ is of class C^{k-j} for any fixed $\delta' < \delta$.

Since $|W|_{0,\delta} \leqslant 2r$, it is clear from (9) that $|U|_{0,\delta} \leqslant C|W|_0^{\ell+1} \leqslant \widetilde{C}r^{\ell+1}$ where C and $\widetilde{C}$ are uniform constants. This estimate together with the estimate on v given in (8) imply

$$|A(r, \lambda, y, w, z)(\zeta)| \leqslant Cr_0$$

where $C = C(\delta)$ is a constant independent of $(r, \lambda, y, w, z) \in T_{r_0}\times\overline{D}$. Therefore, if $\omega \ni 0$ is the open set in M given in Theorem II, we may further restrict r_0 so that boundary $\{A(r, \lambda, y, w, z)(\cdot)\} \subset \omega$ for $(r, \lambda, y, w, z) \in T_{r_0}\times\overline{D}$.

Now define the map $F_r\colon \mathbb{R}\times\mathbb{R}^d\times\mathbb{C}^m \longmapsto \mathbb{C}^{d+m}$ by

$$F_r(\lambda, y, w) \;=\; A(r, \lambda, y, w, z) = -\lambda(\zeta = -\lambda)$$

We claim that for each fixed $r \leqslant r_0$, F_r is of class C^k for $0 \leqslant \lambda \leqslant 1$, $|y| < r$, $|w| < r$. To see this, we note that a typical term of a *kth*-order derivative of F_r involves a derivative of A of order $k-j$ with respect to the parameters λ, y, w, z and a *jth*-order derivative of A with respect to ζ, $0 \leqslant j \leqslant k$. Since the map

$$(r, \lambda, y, w, z) \longmapsto A(r, \lambda, y, w, z)(\cdot) \in C^{j,\delta'}(\overline{D})$$

is uniformly C^{k-j} for $(r, \lambda, y, w, z) \in T_{r_0}\times\overline{D}$, clearly F_r is of class C^k as claimed.

We let $\widetilde{\omega}_{r,a} = F_r(\lambda, y, w) \in \mathbb{C}^{d+m}$; $a < \lambda \leqslant 1$, $|y| < r$, $|w| < r$. For r fixed with $r \leqslant r_0$ and for a chosen sufficiently close to 1, we will show $\widetilde{\omega}_{r,a}$ is the desired manifold with boundary for Theorem II. Note that each point in $\widetilde{\omega}_{r,a}$ is of the form $F_r(\lambda, y, w) \in A(r, \lambda, y, w, z = -\lambda)(\zeta = -\lambda)$ and boundary $\{A(r, \lambda, y, w, z = -\lambda)(\cdot)\} \subset \omega$ for $(r, \lambda, y, w, z) \in T_{r_0}\times\overline{D}$. Thus Part ii) of Theorem II is ensured. Note that a naive dimension count of the parameters indicates that $\dim_{\mathbb{R}} \widetilde{\omega}_{r,a} = d+2m+1 = \dim_{\mathbb{R}} M+1$. To finish the proof of Theorem II in this case we must show the following.

(I) The set $\{F_r(\lambda = 1, y, w);\ |y| < r,\ |w| < r\}$ is an open subset of M containing 0.

(II)
$$(D_{\lambda,y,w}F_r)(\lambda = 1, y = 0, w = 0) \;=\; \begin{pmatrix} * & 0 & 0 \\ 0 & I_d & 0 \\ 0 & 0 & I_{2m} \end{pmatrix}$$

where $D_{\lambda,y,w}F_r$ is the λ, y, w-*real* Jacobian of F_r and I_j is the $j\times j$-identity matrix and where * indicates some nonzero vector.

Statement (II) above shows that the rank of $(D_{\lambda,y,w}F_r)(\lambda = 1, y = 0, w = 0)$ is maximal $(d+2m+1)$. So $\widetilde{\omega}_{r,a}$ is a C^k-manifold with boundary for a suitably close to 1. Statement (I) states that the (manifold) boundary of $\widetilde{\omega}_{r,a}$ (which corresponds to $\lambda = 1$) is an open subset of M containing 0.

This is Requirement i) in the theorem.

To see (I), we note that $W(r, \lambda, w)(\zeta = -\lambda) = w$. From (5), we have $V(r, \lambda, y, w, z = -\lambda)(\zeta = -\lambda) = y$. So

$$F_r(\lambda, y, w) = A(r, \lambda, w, z = -\lambda)(\zeta = -\lambda) = \begin{pmatrix} U(r, \lambda, y, w, z = -\lambda)(\zeta = -\lambda) + iy \\ w \end{pmatrix}$$

Now let us examine the term.

$$f_r(\lambda, y, w) := U(r, \lambda, y, w, z = -\lambda)(\zeta = -\lambda)$$

Letting $\lambda \longrightarrow 1$ and using (9) we obtain $f_r(\lambda = 1, y, w) = h(y, w)$. So $F_r(\lambda = 1, y, w) = (h(y, w) + iy, w)$, which parametrizes an open subset of M containing 0, for $|y| < r$, $|w| < r$.

Since $Dh(0) = 0$, we will show (II) by establishing

$$\frac{\partial f_r}{\partial \lambda}(\lambda = 1, y = 0, w = 0) = \frac{r^{\ell+1}}{2^{\ell+1}} \nu$$

where ν is the vector in the image of L_0 (cf. (6)). Now $U(r, \lambda, y, w, z)(\zeta)$ is harmonic in ζ, $|\zeta| \leqslant 1$, and therefore we can reproduce $U(r, \lambda, w, z)(\zeta)$ by the Poisson kernel. So

$$\begin{aligned} f_r(\lambda, y, w) &= U(r, \lambda, y, w, z = -\lambda)(\zeta = -\lambda) \\ &= P(\zeta = -\lambda, U(r, \lambda, y, w, z = -\lambda)(\cdot)) \\ &= P(\zeta = -\lambda, H(W(r, \lambda, w))(\cdot)) \end{aligned}$$

where the last equation uses (9). Now, from a Taylor expansion of h we have

$$h(w) = \sum_{|\alpha| + |\beta| = \ell + 1} \frac{1}{\alpha! \beta!} \frac{\partial^{\ell+1} h}{\partial w^\alpha \partial \bar{w}^\beta}(0)\, w^\alpha \bar{w}^\beta.$$

Therefore, we must show

$$(10) \qquad \frac{\partial}{\partial \lambda}\{P(z = -\lambda, \sum_{|\alpha| + |\beta| = \ell+1} \frac{\partial^{\ell+1} h}{\partial w^\alpha \partial \bar{w}^\beta}(0)\, W^\alpha \overline{W}^\beta)\}\,|_{\lambda = 1, y = 0, w = 0} = \frac{-r^{\ell+1}}{2^{\ell+1}} \nu.$$

Note from the formula for $W = (W^1, \ldots, W^m)$ that when $w = 0$, $W^1 = \dfrac{r(\lambda + \zeta)}{1 + \lambda}$ and $W^2 = \ldots = W^m = 0$. So the left side of (10) becomes

$$\frac{\partial}{\partial \lambda}\{(1-\lambda) \sum_{\substack{p+q = \ell-1 \\ p = 0, 1, \ldots, \ell-1}} \frac{r^{\ell+1}(1+\lambda)^{-\ell-1} \partial^{\ell+1} h(0)}{(p+1)!\,(q+1)!\, \partial w_1^{p+1} \partial \bar{w}_1^{q+1}} \cdot \frac{1}{2\pi i} \int_{|\zeta| = 1} (\lambda + \zeta)^{p+1} (\lambda + \bar{\zeta})^{q+1} \left(\frac{1}{\zeta + \lambda} - \frac{1}{1 + \lambda\zeta}\right) d\zeta\}|_{\lambda = 1}$$

Here we have used the following formula for the Poisson kernel

$$P(z,u) = \frac{1}{2\pi i} \int_{|\zeta|=1} u(\zeta)\left[\frac{1}{(\zeta+\lambda)} - \frac{\lambda}{(1+\lambda\zeta)}\right] d\zeta$$

Thus the left side of (10) becomes

$$\frac{\partial}{\partial\lambda}(1-\lambda) \sum_{\substack{p+q=\ell-1 \\ p=0,\dots,\ell-1}} \frac{r^{\ell+1}(1+\lambda)^{-1}}{(p+1)!\,(q+1)!} \frac{\partial^{\ell+1} h(0)}{\partial w_1^{p+1}\, \partial \bar{w}_1^{q+1}} \cdot \frac{1}{2\pi i} \int_{|\zeta|=1} \frac{(\zeta+\lambda)^p (1+\lambda\zeta)^q}{\zeta^{q+1}} d\zeta \Big|_{\lambda=1}$$

The only nontrivial term occurs when $\frac{\partial}{\partial\lambda}$ is applied to $(1-\lambda)$. In addition, the integral can be evaluated using residue theory. Since $\lambda = 1$ we have

$$\frac{1}{2\pi i} \int_{|\zeta|=1} \frac{(\zeta+\lambda)^p (1+\lambda\zeta)^q}{\zeta^{q+1}} d\zeta \Big|_{\lambda=1} = \frac{1}{q!} \frac{d^q}{d\zeta^q} (\zeta+1)^{p+q} \Big|_{\zeta=0} = \binom{\ell-1}{p}$$

since $p+q = \ell-1$. Therefore, the left side of (10) becomes

$$\frac{-r^{\ell+1}}{2^\ell} \sum_{\substack{p+q=\ell-1 \\ p=0,\dots,\ell-1}} \binom{\ell-1}{p} \frac{1}{(p+1)!\,(q+1)!} \frac{\partial^{\ell+1} h(0)}{\partial w_1^{p+1}\, \partial \bar{w}_1^{q+1}} = \frac{-r^{\ell+1}}{2^\ell} \nu$$

by (6). Note that the negative sign on the right side is present because the boundary of $\tilde{\omega}_{r,a}$ corresponds to $\lambda = 1$ and the interior of $\tilde{\omega}_{r,a}$ corresponds to $\lambda < 1$. Thus the orientation is reversed from the usual way of parameterizing a manifold with boundary.

REFERENCES

[AB] A. Boggess, *One-Side CR Extension,* To Appear in **Mich. Math. J.**

[B] E. Bishop, *Differentiable Manifolds in Complex Euclidean Space,* **Duke Math. J.** 32 (1965), 1-22.

[BP] A. Boggess and J. Pitts, *CR Extension Near A Point of Higher Type,* To Appear.

[BPo] A. Boggess and J. C. Polking, *Holomorphic Extension of CR Functions,* **Duke Math. J.** 49 (1982), 757-784.

[BT1] M. S. Baouendi and F. Treves, *A Property of the Functions and Distributions Annihilated by a Locally-Integrable System of Complex Vector Fields,* **Ann. of Math.** 113 (1981), 387-421.

[BT2] M. S. Baouendi and F. Treves, *About the Holomorphic Extension of CR Functions of Real Hypersurfaces in Complex Space,* Preprint.

[BCT] M. S. Baouendi, C. H. Chang, and F. Treves, *Microlocal Hypoanalyticity and Extension of CR Functions,* To Appear.

[Gr] S. J. Greenfield, *Cauchy-Riemann Equations in Several Variables,* **Ann. Scuola Norm. Sup. Pisa Sci. Fis. Mat** 22 (1968), 275-314.

[HT] C. D. Hill and G. Taiani, *Families of Analytic Discs with Boundaries in a Prescribed CR Submanifold,* **Ann. Scuola Norm. Sup. Pisa** 4-5 (1978), 327-380.

[HW] L. R. Hunt and R. O. Wells Jr., *Extensions of CR-Functions,* **Amer. J. Math.** 98 (1976), 805-820.

Texas A&M University
College Station
Texas 78743
U.S.A.

L BRAMBILA

Endomorphisms of vector bundles over a compact Riemann surface

§1. Introduction

In this paper we consider a moduli problem, namely, the classification of nilpotent endomorphisms of indecomposable semistable vector bundles of rank 2 over a compact connected Riemann surface X. We assign to each nilpotent endomorphism a canonical extension

$$0 \longrightarrow L \longrightarrow E \longrightarrow L' \longrightarrow 0.$$

By means of this extension we classify the indecomposable semistable vector bundles of rank 2. Using the universal property of the Picard variety and the universal family of extensions given by Narasimhan and Ramanan in [4], we construct a universal family of endomorphisms for P(2, d). We also give a universal family for isomorphism classes of indecomposable semistable vector bundles of rank 2 which are not simple. In this case the algebra of endomorphisms of the vector bundles varies as a function of the variety that parametrises such bundles. I thank the referee for his comments.

§2. General Results

Let X be a compact connected Riemann surface and let $E \xrightarrow{P} X$ be a holomorphic vector bundle over X. Let us recall that an endomorphism $\phi\colon E \longrightarrow E$ of E is a holomorphic fibre map which induces in each fibre $\phi_x\colon E_x \longrightarrow E_x$ a homomorphism of vector spaces. The space of all endomorphisms of a vector bundle E, denoted by END(E), is a finite-dimensional vector space which is in one-to-one correspondence with the sections of the vector bundle End(E), i.e.,

$$\Gamma(\mathrm{End}(E)) \cong \mathrm{END}(E).$$

Atiyah [1] defined a finite-dimensional algebra to be special if

i) A has a unit element I

ii) The nilpotent elements in A form a subalgebra N

iii) A is a vector space $A \cong (I) \oplus N$, where (I) is the subspace of dimension one generated by I.

It then follows that a vector bundle over X is indecomposable iff END(E) is a special algebra. Thus, for an indecomposable vector bundle E over X any endomorphism ϕ of E can be written as a sum of scalar multiples of the identity and a nilpotent endomorphism. Hence, to study the kind of endomorphisms that a vector bundle can have we can concentrate in studying what kind of nilpotent endomorphisms the vector bundle can have.

Indecomposable vector bundles that have no non-zero nilpotent endomorphisms are called simple. Stable vector bundles are simple. Narasimhan and Seshadri in [3] had already parametrised these vector bundles. They proved that the set of stable vector bundles over X of fixed rank and degree has a natural complex structure. Seshadri, in [6], proved that the set of S-equivalence classes of semistable vector bundles of rank n and Chern class d is a complex manifold.

Here we concentrate our study of semistable vector bundles which are not simple.

We consider the pairs (E, ϕ), where E is an n-dimensional indecomposable semistable vector bundle over X and ϕ is a non-zero nilpotent endomorphism of E. An isomorphism from (E, ϕ) to (F, ψ) is an isomorphism of vector bundles:

$$\alpha: E \longrightarrow F \quad \text{such that} \quad \alpha\phi = \psi\alpha$$

We are concerned with the classification of the pairs (E, ϕ), up to isomorphisms. Denote by $P(n, d)$ the set of isomorphic classes of pairs $[E, \phi]$, where E has rank n and slope d (the slope of E, denoted by $\mu(E)$, is $c(E)/rk(E)$).

Mumford and Suominin [5] introduced the concept of a family of endomorphisms of vector spaces. Using similar definitions we introduce the concept of a universal family of endomorphisms of vector bundles.

Definition 1. *A global family of endomorphisms of n-dimensional vector bundles over* X *parametrised by a variety* T *is a pair* (W, Φ), *where* W *is a vector bundle of rank* n *over* $X \times T$ *and* Φ *is an endomorphism of* W. *The family* (W, Φ) *is called* universal *if for every family* (V, Ψ) *parametrised by a variety* S *there is a unique morphism*

$$\theta: S \longrightarrow T \quad \text{with} \quad \theta^*(W, \Phi)_{\theta(s)} \sim (V, \Psi)_s$$

As a first attempt in search of a moduli space for $P(n, d)$ it is natural to ask if there exists a family of endomorphisms in which each isomorphism class is represented exactly once, and if it exists, when it is universal. In general this family does not exist. However, we can restrict the class of isomorphisms suitably to obtain a moduli space.

§3. Rank 2 Case

The situation for $P(2, d)$ is very special. In this case we have that for any indecomposable semistable vector bundle of rank 2, its algebra of endomorphism has dimension less than or equal to 2. Hence, any non-zero nilpotent endomorphism has index 2.

We will show that in this case there is a universal family of endomorphisms such that each element of $P(2, d)$ is represented exactly once.

Let $[E, \phi]$ be a pair in $P(2, d)$. Let us recall that over a compact connected Riemann surface every torsion-free sheaf is locally-free. Hence, we see that the kernel and image of ϕ define subbundles K_ϕ and I_ϕ of E.

Using the canonical factorisation of a homomorphism (see [3]) we have the following exact sequence

$$(1) \qquad 0 \longrightarrow K_\phi \longrightarrow E \longrightarrow I_\phi \longrightarrow 0$$

of vector bundles over X.

Since $\phi^2 = 0$, we see that $I_\phi \subseteq K_\phi$. Both being line bundles, they are the same which we denote by $L \cong K_\phi = I_\phi$.

Thus, given a pair $[E, \phi]$, there is a non-trivial extension

$$\zeta: 0 \longrightarrow L \xrightarrow{i} E \xrightarrow{p} L \longrightarrow 0$$

of a line bundle L by L, with $(E, \phi) \sim (E, (i \circ p))$.

Moreover, if L is a line bundle with $c(L) = d$, then for any non-trivial extension

$$\zeta: 0 \longrightarrow L \xrightarrow{i} F \xrightarrow{p} L \longrightarrow 0$$

of L by L we have

i) F is indecomposable and semistable

ii) the endomorphism $\phi: F \longrightarrow F$ defined as the composition $(i \circ p)$ is nilpotent of index 2.

Hence, we see that a non-trivial extension of a line bundle L by itself, with $c(L) = d$, define a pair $[F, i \circ p]$ in $P(2, d)$.

Let $\mathbf{1}$ be the trivial line bundle over X. Denote by T the affine space $\mathrm{Ext}(\mathbf{1}, \mathbf{1}) \cong \mathbb{C}^g$, where g is the genus of X. By Proposition 3.1 in [4], there is an extension

$$(2) \qquad \zeta: 0 \longrightarrow \mathbf{1}_{X \times T} \xrightarrow{i} \overline{W} \xrightarrow{P} \mathbf{1}_{X \times T} \longrightarrow 0$$

of vector bundles over $X \times T$, where $\overline{W}$ is the universal family of extensions.

If L is the canonical line bundle over $X \times \mathrm{Pic}_d(X)$ then, using the extension (2) we have the following theorem:

Theorem 1. *There is a universal family of endomorphisms* (W, Φ) *parametrised by a variety* M *such that the underlying set is precisely the set* $P(2, d)$.

Proof: Let T° be $\mathrm{Ext}(\mathbf{1}, \mathbf{1}) - 0$ and let

$$(3) \qquad 0 \longrightarrow \mathbf{1}_{X \times T^\circ} \xrightarrow{i} W_\circ \xrightarrow{P} \mathbf{1}_{X \times T^\circ} \longrightarrow 0$$

be the extension (2) restricted to $X \times T^\circ$

If $\qquad \Pi_{13}: X \times T^\circ \times \mathrm{Pic}_d \longrightarrow X \times \mathrm{Pic}_d$

and $\qquad \Pi_{12}: X \times T^\circ \times \mathrm{Pic}_d \longrightarrow X \times T^\circ$

are the natural projections then we have the following exact sequence

(4) $$0 \longrightarrow \pi_{12}^*\mathbf{1}_{X\times T^\circ}\otimes\pi_{13}^*L \xrightarrow{\ i\ } \pi_{13}^*W_\circ\otimes\pi_{13}^*L \xrightarrow{\ P\ } \pi_{12}^*\mathbf{1}_{X\times T^\circ}\otimes\pi_{13}^*L \longrightarrow 0$$

of vector bundles over $X\times T^\circ\times Pic_d$, where $i = \pi_{12}^* i\otimes\pi_{13}^* id_L$ and $p = \pi_{12}^* p\otimes\pi_{13}^* id_L$.

If $M = T^\circ\times Pic_d$ then denote by (W, Φ) the pair $(\pi_{12}^*W_\circ\otimes\pi_{13}^*L, i\circ p)$ over $X\times M$. The universal properties of T° and Pic_d show that the family of endomorphisms (W, Φ) parametrised by M is the universal family of endomorphisms. To prove that the underlying set of M is P(2, d), take any $m\in M$. We get a non-trivial extension ζ_m: $0\longrightarrow \mathbf{1}\xrightarrow{\ i\ } E\xrightarrow{\ p\ } 1\longrightarrow 0$ and a line bundle L_m such that the exact sequence $\zeta_m\otimes L_m$: $0\longrightarrow L\longrightarrow E\otimes L\xrightarrow{\ p\ } L\longrightarrow 0$ define a unique pair $[E\otimes L, i\circ p]$ in P(2, d).

We shall prove now that there exists a universal family of vector bundles in which the algebra of endomorphisms of the vector bundles varies as a function of the moduli space.

Recall that the extensions in $T = Ext(\mathbf{1}, \mathbf{1})$ define two isomorphic vector bundles iff they are in the same one dimensional subspace of T. So we take $\mathbb{P}(T)$. Using similar arguments of Theorem 1 and Lemma 2.3 of [5] we have the following theorem.

Theorem 2. *There is a universal family $\overline{W} = \{W_m : m\in M\}$ of indecomposable semistable vector bundles of rank 2 and Chern class* d *which are no simple, parametrised by $M = \mathbb{P}(T)\times Pic_d(X)$ such that the algebra of endomorphisms of the vector bundles varies as a function of M.*

Proof. We only have to show that the algebra of endomorphisms of the family given in Lemma 2.3 of [5] varies as a function of $M = \mathbb{P}(T)\times Pic_d(x)$.

For any $m\in M$, W_m is a non-trivial extension $0\longrightarrow L\longrightarrow W_m\longrightarrow L\longrightarrow 0$ of a line bundle L by itself, hence W_m is indecomposable semistable (but non-simple) vector bundle, so

$$END(W_m) \cong \mathbb{C}[t]/_{(t^2)},$$

for any $m\in M$.

To conclude this section we consider the simple vector bundles.

Denote by $U_s(2, d)$ the set of isomorphism classes of simple semistable vector bundles of rank 2 and Chern class d over X. For any E in $U_s(2, d)$ there is an extension $0\longrightarrow L\longrightarrow E\longrightarrow L'\longrightarrow 0$ of two non-isomorphic line bundles with $c(L) = c(L')$. Using this extension we give the universal family of simple vector bundles of rank 2 as follows.

Denote by $\Delta^c Pic_d$ the variety $Pic_d\times Pic_d - \Delta$, where Δ is the diagonal subvariety of $Pic_d\times Pic_d$. If π_{1i}: $X\times Pic_d\times Pic_d\longrightarrow X\times Pic_d$ is the natural projection, then denote by L_i the pull back of the canonical line bundle L over $X\times Pic_d$ under π_{1i}.

For any $\overline{\ell} = (\ell, \ell')$ in $\Delta^c Pic_d$, we have that if $L\in\ell$ and $L'\in\ell'$ then

(5) $$H^0(X, Hom(L', L)) = 0$$

and from the Riemann-Roch Theorem we have that

(6) $$\dim H^1(X, \mathrm{Hom}(L', L)) = g-1.$$

Denote by π: $X \times \Delta^c \mathrm{Pic}_d \longrightarrow \Delta^c \mathrm{Pic}_d$ the projection.

From the inequalities (5) and (6) we have that the direct image $R^{\circ}_{\pi}(L_2^* \otimes L_1)$ is the zero-bundle and if π: $V \longrightarrow \Delta^c \mathrm{Pic}_d$ is the first direct image of $\mathrm{Hom}(L_2, L_1)$ by π then V is a vector bundle of rank $g-1$ over $\Delta^c \mathrm{Pic}_d$.

If $\mathbb{P}(V)$ is the projective bundle associated to the vector bundle V then from Lemma 2.3 in [5] there is a family $U = \{U_p : p \in \mathbb{P}(V)\}$ of vector bundles over X parametrised by $\mathbb{P}(V)$ such that for each $p \in \mathbb{P}(V)$, U_p is isomorphic to the vector bundle obtained as an extension of two non-isomorphic line bundles L and L' such that if $L \in \ell$ and $L' \in \ell'$ then $\pi(p) = (\ell, \ell')$. Since $L \not\cong L'$, U_p is simple.

Hence, the vector bundle U over $X \times \mathbb{P}(V)$ is the universal family of the simple vector bundles of rank 2 and Chern class d when $g \geqslant 2$.

§4. Remarks

In the rank 2-case we could find a moduli space for the set $P(2, d)$ because *all* the non-zero nilpotent endomorphisms have index 2. However, for $n > 2$ we see that there could be $(n-1)$ different kinds of non-zero nilpotent endomorphisms.

For instance, for $n = 3$ the vector bundle could have nilpotent endomorphisms of index 2 or index 3, and so the algebra of endomorphisms will be isomorphic to $\mathbb{C}[r, s]/(r^2, s^2)$, or $\mathbb{C}[t]/_{(t^2)}$ or $\mathbb{C}[t]/_{(t^3)}$. It is not possible to have a moduli space for the whole $P(3, d)$, as we have a jump phenomenon.

However, we can split $P(3, d)$ in subsets that have moduli spaces.

Any pair (E, ϕ) in $P(3, d)$ induces two canonical extensions

$$\xi_0 : 0 \longrightarrow E_2 \longrightarrow E \longrightarrow L = 0$$

$$\zeta_0 : 0 \longrightarrow L \xrightarrow{\ i\ } E_2 \xrightarrow{\ p\ } L' \longrightarrow 0.$$

Fix a non-trivial extension ζ_0: $0 \longrightarrow L \longrightarrow E_2 \longrightarrow L' \longrightarrow 0$ of two line bundles with the same Chern class. Denote by T_{ζ_0} the space $\mathrm{Ext}(L_1, E_2) - \{0\}$. One can show (see Proposition 3 in [4]) that there is a universal family of endomorphisms (U, Ψ) over $X \times T_{\zeta_0}$ that parametrise the pairs (E, ϕ) that have ζ_0 as a canonical extension.

Let $L \not\cong L'$. If U_t is indecomposable then $\mathrm{END}(U_t) \cong \mathbb{C}[t]/_{(t^2)}$ for any $t \in T_{\zeta_0}$. However for $L \cong L'$ we have a jump phenomenon. For any $\zeta \in \mathrm{Ext}(L, L)$ define $R_{\zeta} = \{\xi \in \mathrm{Ext}(L_1 E_2) : p^*(\xi) = \zeta\}$, where $p^*: \mathrm{Ext}(L_0, E_2) \longrightarrow \mathrm{Ext}(L, L)$ is the induced map. Hence,

$$\mathrm{END}(U_t) = \begin{cases} \mathbb{C}[r,s]/_{(r^2\, s^2)} & \text{if} \quad t \in R_0 \\ \mathbb{C}[t]/_{(t^3)} & \text{if} \quad t \in R_{\lambda\zeta_0} \quad \text{for} \quad \lambda \in \mathbb{C}^* \\ \mathbb{C}[t]/_{(t^2)} & \text{if} \quad t \in R_\zeta \quad \text{for} \quad \zeta \neq \lambda\zeta_0 \quad \text{and} \quad \lambda \in \mathbb{C}. \end{cases}$$

REFERENCES

[1]. M. F. Atiyah, *Complex Analytic Connections in Fibre Bundles,* **Trans. Amer. Math. Soc.** 85 (1957), p. 181-207.

[2]. D. Mumford and K. Suominen, *Introduction to the Theory of Moduli,* **5th Nordic Summer School in Math.**, Oslo, August 5-25, (1970), p. 171-222.

[3]. M. S. Narasimhan and C. S. Seshadri, *Stable and Unitary Vector Bundles on A Compact Riemann Surface,* **Ann. of Math.** 82 (1965), p. 540-567

[4]. M. S. Narasimhan and S. Ramanan, *Moduli of Vector Bundles on A Compact Riemann Surface,* **Ann. of Math.** 89 (1969), p. 19-51.

[5]. S. Ramanan, *The Moduli Spaces of Vector Bundles over An Algebraic Curve,*. **Math. Ann.** 200 (1973), p. 69-84.

[6]. C. S. Seshadri, *Space of Unitary Vector Bundles on A Compact Riemann Surface,* **Ann. of Math.** 85 (1967), p. 330-336.

Departamento de Matemáticas
Universidad Autónoma Metropolitana
Iztapalapa, México D.F. 09340
México

R EPHRAIM

A generalized criterion for equisingularity

§1. Introduction

Equisingularity theory was introduced by Zariski [e.g., 7, 8, 9] as an approach to the classification of singularities of analytic varieties. The equisingularity theory of families of plane curve singularities is particularly well developed with a variety of geometric, algebraic, numerical and topological criteria known to be equivalent to each other, this convergence of ideas making this an interesting and powerful theory. In this paper we will prove a new discriminant criterion for the equisingularity of families of plane curve singularities, one that is significantly more general than Zariski's discriminant criterion [9]. We prove:

Theorem 1: *Let* V *be a reduced surface in a domain* U *in* $\mathbb{C}^{n+1}$, W *be a domain in* $\mathbb{C}^n$, *and let* F: V $\longrightarrow$ W *be a finite proper map which is smooth off of* $F^{-1}(\Delta)$, Δ *being a proper submanifold of* W. *Then, either* V *is smooth, or the singular locus* SgV *is smooth of dimension* n–1 *and* V *is equisingular along* SgV.

The discriminant criterion presented by Zariski required an additional hypothesis, namely that F should be locally the restriction of a projection map $\mathbb{C}^{n+1} \longrightarrow \mathbb{C}^n$ (i.e., that for all $p \in V$, $F^{-1}(F(p))$ should be embeddable as a hypersurface in $\mathbb{C}$). Theorem 1 shows this additional hypothesis to be unnecessary. Zariski's proof cannot be modified to prove Theorem 1 since it used the extra hypothesis in an essential way. We will deduce Theorem 1 from:

Theorem 2: *Let* V *be a reduced hypersurface in a domain* U *in* $\mathbb{C}^{n+1}$, *let* W *be a domain* in $\mathbb{C}^n$, *and let* Φ: W $\longrightarrow$ V *be a holomorphic homeomorphism. Then if* p *is a smooth point of* SgV, $\dim_p$ SgV = n–1 *and* V *is equisngular along* SgV *at* p.

The reduction of Theorem 1 to Theorem 2 will use elementary covering space arguments. The proof of Theorem 2 will use the μ-constant (constant Milnor number) criterion of Le-Ramanujam [4].

§2. Reduction of Theorem 1 to Theorem 2

Let U, V, W, F, and Δ be as in the statement of Theorem 1, and suppose $p \in V$ is chosen arbitrarily. To prove Theorem 1 we must show that either

(a) p is a smooth point of V

or

(b) p is a smooth point of SgV, $\dim_p SgV = n-1$, and V is equisingular along SgV at p.

If $F(p) \notin \Delta$, then p is a smooth point of V so we may as well assume $F(p) \in \Delta$. Since the conditions of (a) and (b) above are all local in nature we may replace W by any neighborhood W′ of F(p) provided we also shrink V to $F^{-1}(W')$. We may also replace V by the connected component of V containing p. Finally, note that Δ may be enlarged as long as its enlargement is a proper submanifold of W. Combining these remarks we may choose a local coordinate system $(z_1, \ldots, z_n)$ centered at F(p) and reduce to the case:

(1) W is the unit polydisc

(2) $\Delta = W \cap \{z_n = 0\}$.

(3) $F^{-1}(0) = \{p\}$

(4) V is a bounded subset of $\mathbb{C}^{n+1}$

Then [3] $F\colon V - F^{-1}(\Delta) \longrightarrow W - \Delta$ is a finitely-sheeted covering map and $Sg(V) \subset F^{-1}(\Delta)$. To prove Theorem 1 it suffices to show that $F^{-1}(\Delta)$ is an n–1 dimensional manifold which is contained in every irreducible component of V, and that each such irreducible component is either smooth or it has $F^{-1}(\Delta)$ as its singular locus in which case it is equisingular along $F^{-1}(\Delta)$.

Let M be any connected component of $V - F^{-1}(\Delta)$. Then $F|_M\colon M \longrightarrow W - \Delta$ is a connected k-fold covering map. But the fundamental group of $W - \Delta$ is $\mathbb{Z}$, so $F|_M$ is equivalent to the covering map $\eta\colon W - \Delta \longrightarrow W - \Delta$ defined by $\eta(z_1, \ldots, z_n) = (z_1, \ldots, z_{n-1}, z_n^k)$. Let $\varphi\colon W - \Delta \longrightarrow M$ be a covering equivalence for these covering maps. φ is a biholomorphic equivalence. Letting $j\colon M \longrightarrow \mathbb{C}^{n+1}$ be the inclusion, $j \circ \varphi\colon W - \Delta \longrightarrow \mathbb{C}^{n+1}$ is a bounded holomorphic map. By the Riemann Extension Theorem, this extends to a holomorphic map defined on W whose image is easily seen to be contained in V (using the properness of F and the formula $F \circ \varphi(z_1, \ldots, z_n) = (z_1, \ldots, z_{n-1}, z_n^k)$). Thus φ extends to $\tilde{\varphi}\colon W \longrightarrow V$ with $F \circ \tilde{\varphi}\colon W \longrightarrow W$ given by $F \circ \tilde{\varphi}(z_1, \ldots, z_n) = (z_1, \ldots, z_{n-1}, z_n^k)$ which is proper. Thus $\tilde{\varphi}$ is proper and $\tilde{\varphi}(W)$ is an irreducible component of V. Since $F \circ \tilde{\varphi}$ is the identity on Δ, $\tilde{\varphi}(\Delta)$ is an n–1-dimensional submanifold contained in $F^{-1}(\Delta)$; in fact $\tilde{\varphi}(\Delta) = F^{-1}(\Delta) \cap \tilde{\varphi}(W)$. In addition $\tilde{\varphi}$ is injective. (It is injective on $W - \Delta$ and on Δ and $\tilde{\varphi}(W - \Delta) \cap \tilde{\varphi}(\Delta) = \emptyset$ since $\tilde{\varphi}(W - \Delta) \subset F^{-1}(W - \Delta)$ and $\tilde{\varphi}(\Delta) \subset F^{-1}(\Delta)$). If $\Phi\colon W \longrightarrow \tilde{\varphi}(W)$ is the map defined by $\tilde{\varphi}$ then Φ is thus a holomorphic homeomorphism.

If we write $F^{-1}(W - \Delta) = \bigcup_{i=1}^{\ell} M_i$, as the union of its connected components, the analysis of the previous paragraph gives for each i a holomorphic homeomorphism $\Phi_i\colon W \longrightarrow V_i = \tilde{\varphi}_i(W)$ where V_i is an irreducible component of V, and where $V = \bigcup_{i=1}^{\ell} V_i$. For each i, $\Phi_i(\Delta) = F^{-1}(\Delta) \cap V_i$ is an n–1-dimensional manifold. We want to show that in fact $F^{-1}(\Delta) = F^{-1}(\Delta) \cap V_i$ for each i. If V is irreducible this is immediate. If not, we must show that for $i \neq i'$, $\Phi_i(\Delta) = \Phi_{i'}(\Delta)$. But $V_i \cap V_{i'} \subset Sg(V)$

$\subset F^{-1}(\Delta)$. Thus $V_i \cap V_{i'} \subset V_i \cap F^{-1}(\Delta) = \Phi_i(\Delta)$, since V_i and $V_{i'}$ are hypersurfaces $\dim V_i \cap V_{i'} = n-1$. Since $\Phi_i(\Delta)$ is a connected n–1-dimensional manifold $V_i \cap V_{i'} = \Phi_i(\Delta)$. Similarly $V_i \cap V_{i'} = \Phi_{i'}(\Delta)$ and $\Phi_{i'}(\Delta) = \Phi_i(\Delta)$. Thus $F^{-1}(\Delta)$ is a connected n–1-dimensional manifold contained in every irreducible component of V.

The proof of Theorem 1 would be complete if it could be shown that for each i either V_i is smooth, or $\mathrm{Sg}\,V_i = F^{-1}(\Delta)$ and V_i is equisingular along $F^{-1}(\Delta)$. Theorem 2 is applied to the holomorphic homeomorphism $\Phi_i: W \longrightarrow V_i$ shows that if V_i is not smooth, then $\mathrm{Sg}\,V_i$ is of pure dimension n–1. Since $\mathrm{Sg}\,V_i \subset \mathrm{Sg}\,V \subset F^{-1}(\Delta)$ either V_i is smooth or $\mathrm{Sg}\,V_i = F^{-1}(\Delta)$, and, in the latter case, another application of Theorem 2 gives V_i equisingular along $F^{-1}(\Delta)$.

This completes the reduction of Theorem 1 to Theorem 2.

§3. Proof of Theorem 2

Now let V, U, W, Φ, and p be as in the statement of Theorem 2. Let $\Phi^{-1}: V \longrightarrow W$ be the inverse homeomorphism. Φ^{-1} is certainly holomorphic at all smooth points of V. But then, being continuous, it must be holomorphic at all normal points of V. Since W is smooth, every normal point of V is actually a smooth point of V. Thus $\mathrm{Sg}\,V$ consists entirely of non-normal points. Since V is a hypersurface, it follows from a result of Oka [6] that $\mathrm{Sg}\,V$ is of pure dimension n–1. This proves the first assertion of Theorem 2.

We must now show that V is equisingular along $\mathrm{Sg}\,V$ at p. If $n=1$ then V is just a plane curve and there is nothing to show. Since the result to be shown is local in nature, we may simultaneously shrink V about p and W about $\Phi^{-1}(p)$. Φ restricts to a holomorphic homeomorphism $\Phi^{-1}(\mathrm{Sg}\,V) \longrightarrow \mathrm{Sg}\,V$. Shrinking V and W we may assume $\mathrm{Sg}\,V$ is smooth. The map $\Phi^{-1}(\mathrm{Sg}\,V) \longrightarrow \mathrm{Sg}\,V$ is then a biholomorphic equivalence so $\Phi^{-1}(\mathrm{Sg}\,V)$ is an n–1-dimensional submanifold of W. Choosing local coordinates about p and further shrinking we can assume $\mathrm{Sg}\,V = V \cap \{z_n = z_{n+1} = 0\}$. Then $(z_1, \ldots, z_{n-1})$ restrict to a coordinate system of $\mathrm{Sg}\,V$, so $(z_1 \circ \Phi, \ldots, z_{n-1} \circ \Phi)$ restrict to a coordinate system for $\Phi^{-1}(\mathrm{Sg}\,V)$. $(z_1 \circ \Phi, \ldots, z_{n-1} \circ \Phi)$ together with a local equation for $\Phi^{-1}(\mathrm{Sg}\,V)$ thus form a local coordinate system near $\Phi^{-1}(p)$. Shrinking still further, we finally may assume:

(1) $W = W_1 \times W_2$ where W_1 is a ball in $\mathbb{C}^{n-1}$ centered at O having coordinates $s = (s_1, \ldots, s_{n-1})$ and W_2 is a disc in $\mathbb{C}$ centered at O having coordinate t.

(2) V is a reduced hypersurface in a domain U in $W_1 \times \mathbb{C}^2$; the factor $\mathbb{C}^2$ having coordinates (x, y).

(3) $\mathrm{Sg}\,V = W_1 \times \{0\}$

(4) V has a global defining equation $f(s, x, y) = 0$ in U.

(5) The holomorphic homeomorphism $\Phi: W \longrightarrow V$ is defined by $\Phi(s, t) = (s, x(s, t), y(x, t))$ with $x(s, 0) = y(s, 0) = 0$.

We will assume for the rest of this section that the above reduction has been made.

For $a = a(a_1, \ldots, a_{n-1}) \in W_1$, $U_a = \{(x, y) \in \mathbb{C}^2 \mid (a, x, y) \in U\}$ is a domain $\mathbb{C}^2$. We define V_a to be the analytic subspace of U_a defined by $f(a, x, y)$. Then $V_a \cong V \cap (\{a\} \times \mathbb{C}^2)$ (the "scheme theoretic" intersection). But Φ restricts to a holomorphic homeomorphism from $\{a\} \times W_2$ to $[V \cap (\{a\} \times \mathbb{C}^2)]_{red}$, the reduction of $V \cap (\{a\} \times \mathbb{C}^2)$. Thus V_a is topologically a disc so that $f(a, x, y) \not\equiv 0$. Thus V_a is a hypersurface in U_a with $(V_a)_{red}$ irreducible (since it is topologically a disc). Thus, either V_a is reduced or the non-reduced locus of V_a is all of V_a. But the holomorphic homeomorphism $\Phi: W \longrightarrow V$ restricts to a biholomorphic equivalence $W - (W_1 \times \{0\}) = W_1 \times (W_2 - \{0\}) \longrightarrow V - \mathrm{Sg}\, V$. This induces a biholomorphic equivalence between $\{a\} \times (W_2 - \{0\})$ and $V \cap [\, a \times (\mathbb{C}^2 - \{0\})] \cong V_a - \{0\}$. $V_a - \{0\}$, being smooth, is reduced. Thus V_a is reduced.

We have just seen for all $a \in W_1$ V_a is a plane curve which is non-singular except at the origin. Let μ_a be the Milnor number of the singularity of V_a at the origin. By the Le-Ramanujam result [4], to show that V is equisingular along $\mathrm{Sg}\, V = W_1 \times \{0\}$ it suffices to show that μ_a is independent of $a \in W_1$. But to do this it suffices to show that for any complex line L through $0 \in \mathbb{C}^{n-1}$, μ_a is independent of $a \in W_1 \cap L$.

Making a unitary change of coordinates in $\mathbb{C}^{n-1}$ we may suppose $L = \{(s_1, \ldots, s_{n-1}) \mid s_j = 0 \text{ for } j \geqslant 2\}$. Let $W_1' = \{s_1 \in \mathbb{C} \mid (s_1, 0, \ldots, 0) \in W_1 \cap L\}$. Then W_1 is a ball in $\mathbb{C}$ (i.e., a disc) centered at 0 having coordinate s_1. Let U' be the domain in $W_1' \times \mathbb{C}^2$ defined by $U' = \{(s_1, x, y) \mid (s_1, 0, \ldots, x, y) \in U\}$. Let V' be the analytic subspace of U' defined by $f(s_1, 0, \ldots, 0, x, y)$. But $f(s_1, 0, \ldots, x, y) \not\equiv 0$ since in fact $f(0, 0, \ldots, 0, x, y) \not\equiv 0$. Thus V' is a hypersurface in U'. Also $V' \cong V \cap (L \times \mathbb{C}^2)$ (the "scheme theoretic" intersection). But Φ restricts to a holomorphic homeomorphism from $(W_1 \cap L) \times W_2$ to $[V \cap (L \times \mathbb{C}^2)]_{red}$ so that V' is topologically a polydisc. V'_{red} is thus irreducible. Since V' is a hypersurface it follows that either V' is reduced or the non-reduced locus of V' is all of V'. But the biholomorphic equivalence $W_1 \times (W_2 - \{0\}) \longrightarrow V - \mathrm{Sg}\, V$ determined by Φ restricts to a biholomorphic equivalence between $(W_1 \cap L) \times (W_2 - \{0\})$ and $V \cap [L \times (\mathbb{C}^2 - \{0\})] \cong V' - (W_1' \times \{0\})$. Thus, $V' - (W_1' \times \{0\})$ is smooth, so it is reduced. This shows that V' is reduced and that $\mathrm{Sg}\, V' \subset W_1' \times \{0\}$. But we certainly have $W_1' \times \{0\} \subset \mathrm{Sg}\, V'$ since $W_1 \times \{0\} = \mathrm{Sg}\, V$. Thus $\mathrm{Sg}\, V' = W_1' \times \{0\}$.

Now let $W' = W_1' \times W_2$ and define $\Phi': W' \longrightarrow V'$ by $\Phi'(s_1, t) = (s_1, x(s_1, 0, \ldots, 0, t), y(s_1, 0, \ldots, 0, t))$. Then Φ' is a holomorphic homeomorphism. (It is essentially the same as the restriction of Φ from $(W_1 \cap L) \times W_2$ to $V \cap (L \times \mathbb{C}^2)$.)

$\dim V' = 2$, and $W' = W_1' \times W_2$, U', V'; and $\Phi': W' \longrightarrow V'$ already satisfy the conditions (1)–(5) of the standard reduction made above for W, U, V, and Φ. Defining V_α' for $\alpha \in W_1'$ the same way that V_a was defined for $a \in W_1$ we actually have $V_\alpha' = V_a$ for $a = (\alpha, 0, \ldots, 0)$. Thus

$$\{V_\alpha' \mid \alpha \in W_1'\} = \{V_a \mid a \in W_1 \cap L\}.$$

Thus, to show that μ_a is independent of $a \in W_1$ in the general case it suffices to prove this for the case $\dim V = n = 2$.

We now suppose that $n = 2$ and we prove that μ_a is independent of $a \in W_1$. Since V_a is topologically a disc, as was shown above, V_a must have an irreducible singular point at $0 \in V_a$. Thus, the "Milnor formula" [5] for μ_a reduces to $\mu_a = 2\delta_a$. Here $\delta_a = \dim \overline{O}_{V_a,0}/O_{V_a,0}$ where $O_{V_a,0}$ is the local ring of the singularity of V_a at 0 and $\overline{O}_{V_a,0}$ is the local ring of the normalization of the singularity. (This last ring is local rather than semi-local because the singularity is irreducible). To show μ_a is independent of $a \in W_1$ we need to show δ_a is independent of $a \in W_1$.

Let O_W be the structure sheaf of W and O_V be the structure sheaf of V. Since Φ is a homeomorphism $\Phi_*(O_W)$ is a coherent sheaf of O_V algebras containing O_V [2]. Since Φ is biholomorphic off $\Phi^{-1}(\mathrm{Sg}\,V) = W_1 \times \{0\}$ we have the equality of stalks $\Phi_*(O_W)_p = O_{V,p}$ for all $p \in V - \mathrm{Sg}\,V$. Thus $\Phi_*(O_W)/O_V$ is a coherent sheaf of O_V modules supported on $\mathrm{Sg}\,V = W_1 \times \{0\}$. Since W_1 is smooth, the restriction of $\Phi_*(O_W)/O_V$ to $\mathrm{Sg}\,V = W_1 \times \{0\}$ is a coherent sheaf of O_{W_1} modules.

Let $a \in W_1$ be arbitrary, let $p = (a,0) \in \mathrm{Sg}\,V$, and let $q = (a,0) = \Phi^{-1}(p) \in W$. Since $\{a\} \times W_2$ is smooth the holomorphic homeomorphism from $\{a\} \times W_2$ to $V \cap (\{a\} \times \mathbb{C}^2)$ defined by Φ must in fact be the normalization of $V \cap (\{a\} \times \mathbb{C}^2) \cong V_a$. Thus, letting $\tilde{s} = s_1 - a$ we have $\delta_a = \dim (O_{W,q}/\tilde{s}\,O_{W,q}/O_{V,p}/\tilde{s}\,O_{V,p})$. But $O_{W,p}/\tilde{s}\,O_{W,p} \cong O_{W,q} \otimes_{O_{W_1,a}} \mathbb{C}$ and $O_{V,p}/\tilde{s}\,O_{V,p} \cong O_{V,p} \otimes_{O_{W_1,a}} \mathbb{C}$, where we have identified $O_{W_1,a}/\tilde{s}\,O_{W_1,a}$ and $\mathbb{C}$. Also, $O_{W,q} \otimes_{O_{W_1,a}} \mathbb{C} / O_{V,p} \otimes_{O_{W_1,a}} \mathbb{C} \cong (O_{W,q}/O_{V,p}) \otimes_{O_{W_1,a}} \mathbb{C}$. (Here the inclusion $O_{V,p} \subset O_{W,q}$ is the map Φ^*). But $O_{W,q}/O_{V,p} \cong (\Phi_*(O_W)/O_V)_p$. Thus, $\delta_a = \dim[(\Phi_*(O_W)/O_V)_p \otimes_{O_{W_1,a}} \mathbb{C}]$ so to show that δ_a is independent of $a \in W_1$ it suffices to show that $\Phi_*(O_W)/O_V|_{\mathrm{Sg}\,V}$ is a locally-free sheaf of O_{W_1} modules. To do this it suffices to show that $(\Phi_*(O_W)/O_V)_p$ is a free $O_{W_1,a}$-module for any $a \in W_1$. But $O_{W_1,a} \cong \mathbb{C}\{\tilde{s}\}$ (the ring of convergent power series in $\tilde{s}$) which is a discrete valuation ring with maximal ideal generated by $\tilde{s}$. Then to show that $(\Phi_*(O_W)/O_V)_p$ is a free $O_{W_1,a}$-module it suffices [1] to show that it is torsion-free. But to show this it suffices to show that the multiplication map $(\Phi_*(O_W)/O_V)_p \xrightarrow{\cdot\tilde{s}} (\Phi_*(O_W)/O_V)_p$ is injective. Suppose $g \in O_{W,q} \cong \Phi_*(O_W)_p$ determines a class in the kernel of this map. Then there is an $h \in O_{V,p}$ satisfying $h \circ \Phi = \tilde{s}g$. Since this function vanishes on $\{a\} \times W_2$ the function h vanishes on $\Phi(\{a\} \times W_2) = V \cap (\{a\} \times \mathbb{C}^2)$. But V_a is reduced, so $V \cap (\{a\} \times \mathbb{C}^2) \cong V_a$ is also reduced. Thus $h \in \tilde{s}\,O_{V,p}$ so $h \circ \Phi = \tilde{s}g \in \Phi^*(\tilde{s}\,O_{V,p}) = \tilde{s}\,\Phi^*(O_{V,p}) \subset O_{W,q}$ so $g \in \Phi^*(O_{V,p})$. Thus g determines the zero class in $(\Phi_*(O_W)/O_V)_p$. Thus, the kernel of the above multiplication is trivial, and we are done.

REFERENCES

[1]. Douady, A.: *"Flatness and Privilege"*, **Enseignement Math.** 14, (1968), pp. 47-74

[2]. Grauert, H.: *"Ein Theorem der Analytischen Garbentheorie und die Modulräume Komplexer Strukturen"*, **Inst. Hautes Etudes Sci. Publ. Math. No.** 5 (1960).

[3]. Gunning, R. and Rossi, H.: *Analytic Functions of Several Complex Variables,* Prentice Hall Inc., Englewood Cliffs, N. J. (1965).

[4]. Le D. T. and Ramanujam, C. P.: *"The Invariance of Milnor's Number Implies the Invariance of the Topological Type"*, **Amer. J. Math.** 98, (1976), pp. 67-78.

[5]. Milnor, J..: *Singular Points of Complex Hypersurfaces,* **Ann. Math. Stud.** 61, Princeton University Press, (1968).

[6]. Oka, K.: *"Sur les fonctiones analytiques de plusieurs variables VIII"*, **J. Math. Soc. Japan** 3, (1951).

[7]. Zariski, O.: *"Equisingular Points on Algebraic Varieties",* **Seminari dell'Instituto Nazionalle di Alta Mathematica,** 1962-1963, Edizioni Cremonese, Roma (1964) pp. 164-177.

[8]. Zariski, O.: *"Studies in Equisingularity I. Equivalent Singularities of Plane Algebroid Curves"*, **Amer. J. Math.** 87. (1965), pp. 507-536.

[9]. Zariski, O.: *"Studies in Equisingularity II. Equisingularity in Co-dimension 1 (and Characteristic Zero)"*, **Amer. J. Math.** 87, (1965), pp. 972-1006.

Department of Mathematics
George Mason University
Fairfax, Virginia 22030
U.S.A.

Currently at:
Bell Communications Research, Inc.
331 Newman Springs Rd.
Red Bank, N.J. 07701
U.S.A.

M A GUEST

Geodesics, harmonic maps and the Yang–Mills equations

This note presents a problem concerning holomorphic curves in compact Kähler manifolds. No new results are proved; the aim is simply to discuss a situation which has occurred recently in the study of the Yang-Mills equations and in the theory of harmonic maps, and by means of which one may compare the behaviour of the Yang-Mills functional with that of the energy functional.

Geodesics: Let $\gamma: S^1 \longrightarrow M$ be a smooth map of the unit circle into a Riemannian manifold M; let ∇ denote the induced covariant derivative in $V = \gamma^{-1}TM$. Sections of V may be considered as "tangent vectors" to the curve γ, where the latter is thought of as a point in the space $\mathrm{Map}(S^1, M)$ of all smooth maps, and one has a "canonical tangent vector" $\gamma' = d\gamma(d/dt)$. The curve γ is a *geodesic* if it satisfies the equation

$$\nabla_{d/dt}\gamma' = 0$$

and this is equivalent to saying that γ is an extremum for the *energy functional*

$$E: \gamma \longmapsto \int_{S^1} |\gamma'|^2.$$

It is a classical fact, due to M. Morse (see [10]), that the space $\mathrm{Map}(S^1, M)$ can be described up to homotopy in terms of its (not necessarily connected) subspace $\mathrm{Geod}(S^1, M)$, of geodesics. For example, in a fixed component $\mathrm{Map}_*(S^1, M)$, the subspace X consisting of minimal energy (i.e., minimal length) geodesics has the following property.

Theorem 1. $\pi_i X \cong \pi_i \mathrm{Map}_*(S^1, M)$, $i < d-1$, *where* d *is the lowest index of any non-minimal geodesic in* $\mathrm{Map}_*(S^1, M)$.

Harmonic Maps: Let $\phi: S^2 \longrightarrow M$ be a smooth map of S^2 into a Kähler manifold M; let D denote the covariant derivative in the complex vector bundle $V = \phi^{-1}TM$ induced from the canonical Hermitian connection of M. The underlying real covariant derivative ∇ is that induced from the Levi-Cività connection of M. Regarding $S^2 \cong \mathbb{CP}^1$ as the one point compactification of $\mathbb{R}^2 \cong \mathbb{C}$, there are local vector fields $\partial/\partial x, \partial/\partial y \in \Gamma_U(TS^2)$ and local complex vector fields $\partial/\partial z = \frac{1}{2}(\partial/\partial x - i\partial/\partial y) \in \Gamma_U(T_{1,0}S^2)$, $\partial/\partial\bar{z} = \frac{1}{2}(\partial/\partial x + i\partial/\partial y) \in \Gamma_U(T_{0,1}S^2)$, where $TS^2 \otimes \mathbb{C} = T_{1,0}S^2 \oplus T_{0,1}S^2$ as usual. Here, Γ_U refers to sections over a suitable open subset U of S^2. With respect to the corresponding decomposition of $T^*S^2 \otimes \mathbb{C}$, $D = D' + D''$ and $D'_{\partial/\partial z} = \frac{1}{2}(\nabla_{\partial/\partial x} - i\nabla_{\partial/\partial y})$, $D''_{\partial/\partial\bar{z}} = \frac{1}{2}(\nabla_{\partial/\partial x} + i\nabla_{\partial/\partial y})$. The "canonical real tangent vectors" to ϕ are $\partial\phi/\partial x = d\phi(\partial/\partial x)$, $\partial\phi/\partial y = d\phi(\partial/\partial y) \in \Gamma_U(V)$, and the corresponding complex vectors are $\partial'\phi = \partial\phi(\partial/\partial z) \in \Gamma_U(V_{1,0})$, $\partial''\phi = \overline{\partial}\phi(\partial/\partial \overline{z}) \in \Gamma_U(V_{0,1})$ where $V \otimes \mathbb{C} = V_{1,0} \oplus V_{0,1}$, and where $\partial\phi: T_{1,0}S^2 \longrightarrow T_{1,0}M$, $\overline{\partial}\phi: T_{0,1}S^2 \longrightarrow T_{1,0}M$ are the usual components of $d\phi \otimes \mathbb{C}$. The map ϕ is *harmonic* if $\mathrm{tr}\nabla d\phi = 0$ (see [2]), a condition which applies equally well to a map ϕ between arbitrary Riemannian manifolds. In our situation, this translates into the equation

$$\nabla_{\partial/\partial x}(\partial\phi/\partial x) + \nabla_{\partial/\partial y}(\partial\phi/\partial y) = 0$$

or, in complex terms,

$$D''_{\partial/\partial z}(\partial'\phi) = 0 \qquad (\text{i.e.,} \quad D'_{\partial/\partial \overline{z}}(\partial''\phi) = 0)$$

which is analogous to the geodesic equation. Indeed, a map is harmonic if and only if it is an extremum for the energy functional

$$E: \phi \longmapsto \int_{S^2} |d\phi|^2.$$

One sees that a holomorphic or anti-holomorphic map is automatically harmonic. It is of interest to know when the converse is true, and the following theorem in J. Eells and J. C. Wood [3] provides an example.

Proposition 2. *Let* $\phi: S^2 \longrightarrow M$ *be harmonic. Assume* (1) ϕ *is an absolute minimum of the energy functional on the subset* $\mathrm{Map}_*(S^2, M)$ *consisting of maps homotopic to* ϕ, *and* (2) $\mathrm{Map}_*(S^2, M)$ *contains at least one holomorphic map. Then* ϕ *is holomorphic.*

This extends to the case $M = F_{r_1, r_2, \ldots, r_m}(\mathbb{C}^n)$ $(= F$, say), i.e., the manifold of complex flags

$$E_{r_1} \subseteq E_{r_2} \subsetneq \ldots \subsetneq E_{r_m}$$

where E_i is a complex linear subspace of $\mathbb{C}^n$ of dimension i (see [6]). The subspace $\mathrm{Hol}_*(S^2, F)$ of $\mathrm{Map}_*(S^2, F)$ is thus the analogue of the space of minimal geodesics. The following analogue of Theorem 1 was proved by the author [6].

Theorem 3. *Let* $r_i = i$, $1 \leqslant i \leqslant m$. *Then* $\pi_i \mathrm{Hol}_*(S^2, F) \cong \pi_i \mathrm{Map}_*(S^2, F)$, $i < d$, *where* d *is an integer depending in a simple way on* m *and the homotopy class of* ϕ.

However, Morse theory is not used in the proof, and the significance of the integer d is not yet clear. This result was based on the paper [12] of G. B. Segal and has since been generalised by F. Kirwan in [8].

The Yang-Mills Equations: Let $P \longrightarrow M$ be a principal bundle with structural group U_n over a compact, oriented Riemannian manifold M, and let $E \longrightarrow M$ be the associated Hermitian vector bundle of rank n. A *connection* in P is a 1-form $\omega \in \Gamma(T^*P \otimes L(U_n))$ satisfying the usual conditions (see [9]), where $L(U_n)$ is the Lie algebra of U_n, and the *curvature form* of ω is the 2-form $\Omega = d\omega + [\omega, \omega] \in \Gamma(\Lambda^2 T^*P \otimes L(U_n))$. It is well known that any vector bundle V associated to $P \longrightarrow M$ via a representation Θ of U_n inherits (from ω) a differential operator

$$D_\Theta \colon \Omega^k V \longrightarrow \Omega^{k+1} V$$

where $\Omega^k V$ denotes the space of V-valued k-forms on M. When V is real we write $\Omega^k V = \Omega^k_{\mathbb{R}} V = \Gamma(\Lambda^k T^*M \otimes V)$, and when V is complex we write $\Omega^k V = \Omega^k_{\mathbb{C}} V = \Gamma(\Lambda^k (T^*M \otimes \mathbb{C}) \otimes V)$. For example, the standard complex representation $\lambda \colon U_n \longrightarrow U_n$ gives rise to $D = D_\lambda \colon \Omega^0_{\mathbb{C}} E \longrightarrow \Omega^1_{\mathbb{C}} E$, and the adjoint representation $\mathrm{Ad} \colon U_n \longrightarrow \mathrm{Aut}(L(U_n))$ gives rise to $D_{\mathrm{Ad}} \colon \Omega^0_{\mathbb{R}} \mathrm{Ad}P \longrightarrow \Omega^1_{\mathbb{R}} \mathrm{Ad}P$ where AdP denotes the associated (real) vector bundle. The connection ω is in fact determined by D, which is the covariant derivative for a connection in E (in the sense of the last paragraph). In these terms, the *curvature operator* is defined to be $D^2 \colon \Omega^0_{\mathbb{C}} E \longrightarrow \Omega^2_{\mathbb{C}} E$; this has order zero and thus defines a form $D^2 \in \Omega^2_{\mathbb{C}} E^* \otimes E$. Since D is "unitary" (i.e., compatible with the Hermitian structure), $(D^2)^{1,1}$ comes from a real form $F \in \Omega^2_{\mathbb{R}} \mathrm{Ad}P$ (note that $E^* \otimes E \cong \mathrm{Ad}P \otimes \mathbb{C}$). This form may be obtained from $\Omega \in \Gamma(\Lambda^2 T^*P \otimes L(U_n))$ by dividing the action of U_n. The *Yang-Mills equations* are:

$$(1) \qquad D^*_{\mathrm{Ad}} F = 0$$

$$(2) \qquad D_{\mathrm{Ad}} F = 0$$

where $D^*_{\mathrm{Ad}} \colon \Omega^2_{\mathbb{R}} \mathrm{Ad}P \longrightarrow \Omega^1_{\mathbb{R}} \mathrm{Ad}P$ is the adjoint of D_{Ad}. These equations say that the connection D has harmonic curvature (actually, (2) is redundant, being the Bianchi identity). Such a connection will be called a *Yang-Mills connection*. The space of connections under consideration is the real affine space associated to the vector space $\Omega^1_{\mathbb{R}} \mathrm{Ad}P$ (it is easy to see that the difference of two connection forms $\omega_1 - \omega_2$ defines an element of this space), and we shall identify it with the vector space by choosing some basepoint. It turns out that a Yang-Mills connection is precisely an extremum for the *Yang-Mills functional* on $\Omega^1_{\mathbb{R}} \mathrm{Ad}P$ defined by the formula $D \longmapsto \int_M |F|^2$. A detailed analysis of Yang-Mills connections when M is a Riemann surface has been given in [1], by utilising the relation between unitary connections and holomorphic structures on E. Since $\dim_{\mathbb{C}} M = 1$ in this case, a holomorphic structure on E is determined by its $\bar\partial$-operator $\bar\partial \colon \Omega^0_{\mathbb{C}} E \longrightarrow \Omega^{0,1}_{\mathbb{C}} E$. The difference of two such operators $\bar\partial_1 - \bar\partial_2$ is of order zero and defines an element of $\Omega^{0,1}_{\mathbb{C}} E^* \otimes E$. Thus, the space of holomorphic structures is the complex affine space associated to the vector space $\Omega^{0,1}_{\mathbb{C}} E^* \otimes E$. We have noted that the underlying real space $\Omega^1_{\mathbb{R}} \mathrm{Ad}P$ parametrises unitary connections on E, and in fact this correspondence is realised by assigning to a covariant derivative D its (0, 1)-component $D'' \colon \Omega^0_{\mathbb{C}} E \longrightarrow \Omega^{0,1}_{\mathbb{C}} E$. This defines a (real) linear map from $\Omega^1_{\mathbb{R}} \mathrm{Ad}P$ to $\Omega^{0,1}_{\mathbb{C}} E^* \otimes E$

on choosing a fixed connection to act as zero. The inverse is given by assigning to a $\bar\partial$-operator $\bar\partial$ the canonical Hermitian connection (whose covariant derivative D, by definition, has the property $D'' = \bar\partial$). Now, $\Omega^1_{\mathbb{R}}\,\mathrm{Ad}P$ admits the action of the group Aut(P) of automorphisms of the principal bundle P, and the Yang-Mills equations are invariant under this action. Similarly, $\Omega^{0,1}_{\mathbb{C}}\,E^*\otimes E$ admits the action of the group $\mathrm{Aut}^{\mathbb{C}}(E)$ of complex automorphisms of E, and the resulting orbits are equivalence classes of holomorphic structures. The action is not free, and it is well known that in order to obtain a good orbit space one must restrict to a subset of $\Omega^{0,1}_{\mathbb{C}}\,E^*\otimes E$, e.g., the *stable* holomorphic structures. Indeed, if the rank of E and its degree deg E are coprime, a theorem of Narasimhan-Seshadri [11] states that the isomorphism classes of stable holomorphic structures form a compact complex manifold (see [1] for further references and more information). A basic result of Atiyah-Bott [1] is the following:

Theorem 4. *Let* E *be an Hermitian bundle over a Riemann surface* M, *such that* rk E *and* deg E *are coprime. Then*

(1) *Under the identification* $\Omega^{0,1}_{\mathbb{C}}\,E^*\otimes E \cong \Omega^1_{\mathbb{R}}\,\mathrm{Ad}P$, *stable holomorphic structures on* E *correspond to minimal Yang-Mills connections (i.e., absolute minima of the Yang-Mills functional).*

(2) *The space of* Aut(P)*-equivalence classes of minimal Yang-Mills connections is isomorphic as a complex manifold to the space of* $\mathrm{Aut}^{\mathbb{C}}(P)$*-equivalence classes of stable holomorphic structures.*

Furthermore, it is shown that an arbitrary Yang-Mills connection arises as a direct sum of minimal Yang-Mills connections $D = D_1 \oplus \ldots \oplus D_r$ for some decomposition $E \cong E_1 \oplus \ldots \oplus E_r$ of E. Thus one has a complete description of the critical manifolds of the Yang-Mills functional on $\Omega^1_{\mathbb{R}}\,\mathrm{Ad}P$, in terms of moduli spaces of holomorphic vector bundles. Technical difficulties have so far prevented a direct attack on the problem of determining whether the Yang-Mills functional on $\Omega^1_{\mathbb{R}}\,\mathrm{Ad}P$ is a "Morse function" (in some generalised sense). However, the above theorem allows one to work in the space $\Omega^{0,1}_{\mathbb{C}}\,E^*\otimes E$ of holomorphic structures, about which more is known. For example, $\Omega^{0,1}_{\mathbb{C}}\,E^*\otimes E$ has a natural decomposition, known as the Harder-Narasimhan stratification, which has the property that each stratum is, up to homotopy, a product of spaces of holomorphic structures on subbundles of E. The main technical result of [1] then asserts that this is an "$\mathrm{Aut}^{\mathbb{C}}(E)$-equivariantly perfect stratification". This gives a relation between moduli spaces of holomorphic structures of subbundles of E, and permits an inductive calculation of their cohomology. It may also be interpreted as saying that the Yang-Mills functional on $\Omega^1_{\mathbb{R}}\,\mathrm{Ad}P$ behaves as a particularly simple type of (generalised) Morse function, at least as far as cohomology is concerned.

In presenting (rather superficially) these three different examples, it has been our aim to emphasize the common goal of obtaining "Morse theoretic" results, and to point out that while this has been done with some success in the first and third cases, the second situation is less well developed.

Results in that case are restricted to the holomorphic (index zero) maps, so there arises a natural question: *do higher index harmonic maps (from a Riemann surface to a Kähler manifold, say) behave in a similar way to higher index critical points in the other two examples?*

One reason for expecting a positive answer to this question is the fact that the third example actually encompasses the other two. If $M = S^2$, Yang-Mills connections correspond to *homogeneous* holomorphic bundles on S^2, which in turn correspond to one parameter subgroups (i.e., geodesics through the identity) in U_n. The theory of the Yang-Mills functional here reduces to that of the energy functional on $\mathrm{Map}(S^1, U_n)$, i.e., we are in the situation of the first example (see [1]). To explain the relation with the second example, we need to give more details of the Harder-Narasimhan stratification of $\Omega^{0,1}_{\mathbb{C}} E^* \otimes E$ mentioned above.

Let E be a holomorphic vector bundle of rank n on a Riemann surface M. Let $\mu(E) = \deg E/\mathrm{rk}\, E$. The bundle E is said to be *stable* (resp. *semistable*) if, for any holomorphic subbundle D of E, $\mu(D) < \mu(E)$ (resp. $\mu(D) \leqslant \mu(E)$). It is known [7] that any E has a canonical filtration $0 = E_0 \subsetneq E_1 \subsetneq \dots \subsetneq E_r = E$ such that:

(1) $D_i = E_i/E_{i-1}$ is semistable

(2) $\mu(D_1) > \mu(D_2) > \dots > \mu(D_r)$

The *type* of E is the n-tuple

$$(\mu(D_1), \dots, \mu(D_1), \mu(D_2), \dots, \mu(D_2), \dots, \mu(D_r), \dots \mu(D_r))$$

where $\mu(D_i)$ is repeated $\mathrm{rk}\, D_i$ times. The bundles of fixed type $\mu = (\mu_1, \dots, \mu_n)$ form a subspace C_μ of the space C of holomorphic bundles (of rank n and degree deg E), and we refer to the decomposition $C = \cup C_\mu$ as the *Harder-Narasimhan stratification.* Of course, $C \cong \Omega^{0,1}_{\mathbb{C}} E^* \otimes E$ as E is determined topologically by its rank and degree. If $M = S^2$, the stratification is easy to describe: E has a holomorphic Grothendieck-Birkhoff splitting $E \cong \sum_{i=1}^{n} L^{a_i}$ as a sum of powers of the standard line bundle L on S^2 which has $c_1 L = 1$, and the type of E is $\mu = (a_1, a_2, \dots, a_n)$.

Let us now return to the consideration of harmonic maps $M \longrightarrow X$, where M is a Riemann surface and X is a compact Kähler manifold. First, it is not surprising that the Harder-Narasimhan stratification may be used to study $\mathrm{Hol}_*(M, X)$. If V is a holomorphic vector bundle on X, we have an inclusion

$$\mathrm{Hol}_*(M, X) \longrightarrow C$$

which associates to a holomorphic map $f: M \longrightarrow X$ of fixed homotopy class the holomorphic bundle $f^{-1}V$. The stratification $C = \cup C_\mu$ induces a stratification of $\mathrm{Hol}_*(M, X)$. In the case $M = S^2$, this was used in [6] to prove the result

$$\pi_i \mathrm{Hol}_*(S^2, F) \cong \pi_i \mathrm{Map}_*(S^2, F), \qquad i < d$$

referred to earlier (Theorem 3), and it was used independently in [8] to prove a similar result. In each case, the method is to proceed by induction, using fibre bundles of the form

$$\mathbb{C}P^{n-a} \longrightarrow F_{a,a+1} \longrightarrow F_a$$

where $F_{i,j,\ldots}$ denotes the space of complex flags $E_i \subseteq E_j \ldots$ of dimensions i, j, ... in $\mathbb{C}^{n+1}$. The main point is to show that the corresponding sequence

$$\mathrm{Hol}_*(S^2, \mathbb{C}P^{n-a}) \longrightarrow \mathrm{Hol}_*(S^2, F_{a,a+1}) \longrightarrow \mathrm{Hol}_*(S^2, F_a)$$

behaves as a fibration after certain strata are removed.

A deeper property of the Harder-Narasimhan stratification is that is exhibits the "lowest index construction" in that a general Yang-Mills connection is given by combining minimal Yang-Mills connections (see earlier remarks); it turns out that the same is true for harmonic maps in certain situations, as we shall explain. The best understood case at the present time is $M = S^2 = \mathbb{C}P^1$, $X = \mathbb{C}P^n$, where all harmonic maps have been described by Eells-Wood [4] in terms of "minimal" harmonic (i.e., holomorphic or antiholomorphic) maps. Let $\phi : \mathbb{C}P^1 \longrightarrow \mathbb{C}P^n$ be a holomorphic map of degree d ($\geqslant 0$). The associated curve $\phi_i : \mathbb{C}P^1 \longrightarrow Gr_{i+1}(\mathbb{C}^{n+1})$ is defined in the usual way [5] by taking at each point the subspace of $\mathbb{C}^{n+1}$ spanned by a local lift of ϕ to $\mathbb{C}^{n+1} - \{0\}$ and its first i derivatives. (This makes sense unless $\phi(\mathbb{C}P^1)$ is contained in a proper subspace $H \cong \mathbb{C}P^r$ with $r < i$; in which case we may define ϕ_i trivially as a suitable constant map.) The maps ϕ_0 $(= \phi)$, $\phi_1, \ldots, \phi_n$ are holomorphic, and define a map $\widetilde{\phi} : \mathbb{C}P^1 \longrightarrow F$, where $F = F_{1,2,\ldots,n}(\mathbb{C}^{n+1})$ now denotes the manifold of full flags in $\mathbb{C}^{n+1}$. The standard ordered basis of $\mathbb{C}^{n+1}$ leads to a natural complex structure on F, with respect to which $\widetilde{\phi}$ is holomorphic. Now, F admits projections $\pi_i : F \longrightarrow \mathbb{C}P^n$, $i = 0, 1, \ldots, n$, given by associating to a flag $\{0\} = E_0 \subseteq E_1 \subseteq \ldots \subseteq E_{n+1} = \mathbb{C}^{n+1}$ the line $E_i^\perp \cap E_{i+1}$ (orthogonal complement refers to the standard Hermitian inner product on $\mathbb{C}^{n+1}$). The result of Eells-Wood [4] may then be stated as follows.

Theorem 5. *The maps* $\pi_i \circ \widetilde{\phi}$ $(i = 0, 1, \ldots, n)$ *are harmonic for any holomorphic map* ϕ. *Moreover, all harmonic maps (from* $\mathbb{C}P^1$ *to* $\mathbb{C}P^n$*) arise through this construction.*

The first statement is proved by a straightforward horizontality argument. To understand the second statement, observe that the maps $\pi_i \circ \widetilde{\phi}$ are obtained from ϕ by taking succesive "holomorphic osculating flags". Given a harmonic map ψ we need a converse process, and a natural candidate is the construction of "antiholomorphic osculating flags", i.e., one where the operator $\partial/\partial z$ is replaced by $\partial/\partial \overline{z}$. This works since a harmonic map of $\mathbb{C}P^1$ into $\mathbb{C}P^n$ is necessarily a branched (minimal) immersion. Starting with ψ itself, it turns out that one obtains a sequence of harmonic maps which terminates with a holomorphic map ϕ, and it follows that $\psi = \pi_i \circ \tilde{\phi}$ for some integer i. (In [4], a version of Theorem 5 is given for maps of an arbitrary Riemann surface into $\mathbb{CP}^n$, but in this case it is no longer true that any harmonic map arises from a holomorphic map by the above construction). In a future article we shall discuss Theorem 5 from a different viewpoint and make some generalisations.

A natural problem is to describe the "moduli space" for each connected component of harmonic maps in terms of their spaces $\text{Hol}_d(\mathbb{CP}^1, \mathbb{CP}^n)$, $d \geqslant 0$. This is not straightforward; for example, if ϕ, $\phi' \in \text{Hol}_d(\mathbb{CP}^1, \mathbb{CP}^n)$ then $\pi_i \circ \tilde{\phi}$, $\pi_i \circ \tilde{\phi}'$ are not in general homotopic. It seems reasonable to expect that the Harder-Narasimhan stratification will play a role here. An estimate for the Morse index of $\pi_i \circ \tilde{\phi}$ is given in [4]. However, there is no possibility of a "classical" application of Morse theory, as the case $n = 1$ shows: here *any* harmonic map of $\mathbb{CP}^1$ to $\mathbb{CP}^1$ (of degree $d \geqslant 0$) is holomorphic (see [2]), yet $\text{Hol}_d(\mathbb{CP}^1, \mathbb{CP}^1)$ is certainly not a retract of $\text{Map}_d(\mathbb{CP}^1, \mathbb{CP}^1)$ as the absence of higher index critical points may suggest.

REFERENCES

[1]. M. F. Atiyah, R. Bott, *"The Yang-Mills Equations over Riemann Surfaces"*, **Phil. Trans. R. Soc. London** A308 (1982), 523-615.

[2]. J. Eells, L. Lemaire, *"A Report on Harmonic Maps"*, **Bull. Lond. Math. Soc.** 10 (1978), 1-68.

[3]. J. Eells, J. C. Wood, *"Maps of Minimum Energy"*, **J. Lond. Math. Soc.** 2 (1981), 303-310.

[4]. J. Eells, J. C. Wood, *"Harmonic Maps from Surfaces to Complex Projective Spaces"*, **Advances in Math.** 49 (1983), 217-263.

[5]. P. Griffiths, J. Harris, *"Principles of Algebraic Geometry"*, Wiley-Interscience (1978).

[6]. M. A. Guest, *"Topology of the Space of Absolute Minima of the Energy Functional"*, **Amer. J. Math.** 106 (1984), 21-42.

[7]. G. Harder, M. S. Narasimhan, *"On the Cohomology Groups of Moduli Spaces of Vector Bundles over Curves"*, **Math. Annalen** 212 (1975), 215-248.

[8]. F. Kirwan, *"On Spaces of Maps from Riemann Surfaces to Grassmannians and Applications to the Cohomology of Moduli of Vector Bundles"*, (to appear).

[9]. S. Kobayashi, K. Nomizu, *"Foundations of Differential Geometry"*, Volumes I, II, Wiley-Interscience (1963, 1969).

[10]. J. W. Milnor, *"Morse Theory"*, **Annals of Math. Studies No.** 51, Princeton University Press (1963).

[11]. M. S. Narasimhan, C. S. Seshadri, *"Stable and Unitary Vector Bundles on a Compact Riemann Surface"*, **Annals of Math.** 82 (1965), 540-547.

[12]. G. B. Segal, *"The Topology of Spaces of Rational Functions"*, **Acta Math.** 143 (1979), 39-72.

Departamento de Matemáticas
Centro de Investigación y de Estudios Avanzados del IPN
México D.F. 07000

Institut des Hautes Etudes Scientifiques
91440 Bures-Sur-Yvette
France

M KALKA

Deformations of submanifolds and vanishing theorems

1. Introduction

The purpose of this note is to give a new proof and extension of the main result of [3]. There we studied the deformation theory of submanifolds of compact Kähler manifolds using the theory of harmonic maps. Here we show how our result follows from general ideas in deformation theory. In particular, our approach is to show that the basis for our theorem are cohomology vanishing theorems. Harmonic maps enter only in that they may be useful in proving these vanishing theorems [4].

We begin by recalling the context of [3]. Let N be a compact complex submanifold of the compact complex manifold M. In this situation one has two deformation space germs of interest. One can study the Kuranishi space, K, of deformations of N as a complex manifold. On the other hand one has the Douday space, D, of deformations of N as a submanifold of M. Our main theorem can now be stated.

Theorem 1. *Suppose* N *is a compact complex submanifold of the compact complex manifold* M. *If* D *and* K *respectively denote the Douady and Kuranishi spaces, then* D *is biholomorphic to* K *if* $H^0(N, TM|N) = H^1(N, TM|N) = 0$.

2. Facts from Deformation Theory

In this section we recall the basic facts from deformation theory that will be required for our proof. A good reference for this material is [1].

Let N be a compact complex manifold and let $f: N \longrightarrow M$ be a holomorphic embedding of N as a complex submanifold of the compact complex manifold M. Let $\mathcal{N}$ denote the normal bundle of the embedding. If D denotes the Douady space of the embedding, then $H^0(N, \mathcal{N})$ is the Zariski tangent space to D. Since the Kuranishi space K of N is complete, there is a holomorphic map $i: D \longrightarrow K$. Our result is that under the conditions of the theorem, i is an isomorphism.

The embedding of $N \longrightarrow M$ gives rise to an exact sequence of holomorphic vector bundles over N:

$$0 \longrightarrow \theta \longrightarrow TM|N \longrightarrow \mathcal{N} \longrightarrow 0.$$

Here θ denotes the holomorphic tangent bundle of N. If we consider the associated long exact sequence in cohomology, we get a coboundary map

$$\delta: H^0(N, \mathcal{N}) \longrightarrow H^1(N, \theta).$$

As is well known, $H^1(N, \theta)$ is the Zariski tangent space to K. What is important for us is that

δ is the map induced on Zariski tangent space by i.

3. Proof of Theorem 1.

The key to the proof of Theorem 1 is the following result.

Lemma. *If* $f: N_0 \longrightarrow M$ *is a holomorphic embedding of compact complex manifolds and*

$$H^1(N_0, TM|N_0) = 0,$$

than there is an extension of f *to the Kuranishi family of* N_0.

The proof of the lemma will rely on techniques of Horikawa [2], together with the following result by Wavrick [5]. In what follows we set $T = TM|N_0$, and we let $\tau: (N, N_0) \longrightarrow (X, 0)$ be a family of compact complex manifolds centered at N_0.

Lemma. *If* $f: N_0 \longrightarrow M$ *extends formally to a map of* $N \longrightarrow M$, *then there is a neighborhood* X_0 *of* 0 *and a holomorphic map* $F: \tau^{-1}(X_0) \longrightarrow M$ *such that* $F|N_0 = f$.

We will now show that if $H^1(N_0, T) = 0$, then f extends formally to N. We assume that X is a curve.

Let N_0 be covered by coordinate neighborhoods $\{U_j\}$ with coordinates z_j and transition functions $z_i = b_{ij}(z_j)$ in $U_i \cap U_j$.

Let $f(N_0)$ be covered by coordinate neighborhoods $\{V_j\}$ in M, with coordinates w_i and transition functions $w_i = g_{ij}(w_j)$ in $V_i \cap V_j$. The map f is given by the functions $w_i = f_i(z_i)$.

Let t represent the coordinate in X. The complex structure on $\tau^{-1}(t)$ is repressented by $\phi(t)$, a (0, 1)-form with values in θ_{N_0}, satisfying

(a) $\bar\partial\phi = 1/2[\phi, \phi]$, $\phi(0) = 0$.

Our extension of f is given by functions Φ_j defined in $U_i \times X$ such that

(b_0) $\Phi_j(z_i, 0) = f_i(z_i)$

(b) $\bar\partial\Phi_i = \phi\cdot\Phi_j$

(c) $\Phi_i(b_{ij}(z_j), t) = g_{ij}(\Phi_j(z_j, t))$.

Let ϕ, Φ_j have Taylor expansions

$$\phi = \sum_{\mu>0} \phi_\mu t^\mu, \quad \Phi_j = \sum_{\mu>0} \Phi_{ij\,\mu} t^\mu$$

with $\Phi_{i,0} = f_i$. Let $\phi^\mu = \phi_1 t + \ldots + \phi_\mu t^\mu$ and $\Phi^\mu_j = \Phi_{i,0} + \ldots + \Phi_{i,\mu} t^\mu$.

Equations (b), (c) are equivalent to the systems

(b′) $\overline{\partial}\Phi_j^\mu = \phi\mu\circ\Phi_j^\mu$

$$\mathrm{mod}\,(t^{\mu+1})$$

(c′) $\Phi_i^\mu(b_{ij}(z_j), t) = g_{ij}(\Phi_j^\mu(z_j, t))$

holding for all μ.

By induction on μ we shall construct functions $\Phi_{i,\mu}$ so that the corresponding $\Phi_i = \Phi_{i,\mu}t^\mu$ satisfy (b′), (c′).

For $\mu = 0$ take $\Phi_{i,\mu} = f_i(z_i)$. Now suppose $\Phi_{i,0}, \ldots, \Phi_{i,\mu-1}$ have been found. Let

$$\xi_\mu = \overline{\partial}\Phi^{\mu-1} - 1/2\,[\Phi^{\mu-1}, \Phi^{\mu-1}]$$

$$-E_{i,\mu} = \overline{\partial}\Phi_j^{\mu-1} - \phi^{\mu-1}\circ\Phi_i^{\mu-1} \qquad \mathrm{mod}\,(t^{\mu-1})$$

$$\Gamma_{i,j,\mu} = \Phi_j^{\mu-1} - g_{ij}(\Phi_j^{\mu-1}).$$

ξ_μ is a global (0, 2)-form with values in θ_{N_0}. The μth term of (a) gives $\overline{\partial}\Phi_\mu = -\xi_\mu$. $E_{i,\mu}$ is a (0, 1)-form with values in T and $\Gamma_{i,j,\mu}$ is a T-valued function on $U_i \cap U_j$. Horikawa verifies the relations below among ξ, E_i, $\Gamma_{i,j}$ (we fix μ for the remainder of the discussion, and thus drop the subscript).

(ii) $\overline{\partial}E_i = f*\xi$

(iii) $\overline{\partial}\Gamma_{i,j} = E_j - E_i$

(iv) $\Gamma_{i,j} + \Gamma_{j,k} + \Gamma_{k,i} = 0$

and equations (b′), (c′) become

(b″) $E_0 = \partial\Phi_i - f*\phi$

(c″) $\Gamma_{i,j} = \Phi_j - \Phi_i$

By (iv) we can find smooth T-valued functions ψ_i defined on U_i such that

$$\Gamma_{i,j} = \psi_i - \psi_j$$

thus, by (iii)

$$\overline{\partial}\psi_i - \overline{\partial}\psi_j = E_j - E_i$$

Thus, $E = E_i + \overline{\partial}\psi_i$ in U_i is a global T-valued (0, 1)-form. By (i), (ii),

$$\overline{\partial}(E+f*\phi) = \partial E_i - f*\xi = 0,$$

so $E+f*\phi$ defines a class in $H^1(N_0, T)$. Since this is zero, there is a T-valued function θ with $\overline{\partial}\theta = E+f*\Phi$. Take $\Phi_i = \theta - \psi_i$. Then (b″) $\overline{\partial}\Phi_i = \overline{\partial}\theta - \overline{\partial}\psi_i = E + f^*\phi - \overline{\partial}\psi_i$

$= E_i + f^*\phi$

(c″) $\Phi_j - \Phi_i = \psi_i - \psi_j = \Gamma_{i,j}$.

Thus, by induction, there is a normal extension. For $\dim X > 1$, replace $\phi_\mu t^m$, $\Phi_{j,\mu} t^\mu$ by homogeneous polynomials of degree μ in $\dim X$ variables. Take these values in the same bundle tensored with the trivial bundle $\otimes^\mu \mathbb{C}^n$. With this modification the same proof goes through.

We now give the proof of Theorem 1. For $x \in N_t$, the tangent space to N_t is $\mathrm{Ker}_x\, d\tau$. For $x \in N_0$, $dF|\mathrm{Ker}_x\, d\tau$ is of maximal rank. Thus, this persists over a neighborhood X' of 0. Furthermore, by the Inverse Function Theorem with parameters, for each $p \in N_0$, there is a neighborhood U_p in N, such that $F|N_t \cap U_p$ is injective.

$F|N_t$ is locally an embedding for $t \in X'$; we have to show that for t near 0, $F|N_t$ is injective. If not, there exists $p_t \neq q_t \in N_t$ such that $F(p_t) = F(q_t)$. As $t \longrightarrow 0$, $p_t \longrightarrow p_0$, $q_t \longrightarrow q_0$. Then $F(p_0) = F(q_0)$, so $p_0 = q_0$. Then, for t close enough to 0, $p_t, q_t \in N_t \cap U_{p_0}$, which is a contradiction. Thus, there exists a neighborhood of 0 on which $F|N_t$ is injective.

We now show that the map $i: D \longrightarrow K$ is a biholomorphism at 0. We note that by cohomology vanishing di is an isomorhism, so $i(D)$ is a subvariety of K. In fact $i(D)$ is a union of irreducible branches of K. Let V be a branch of K. There is an integer μ_0 such that if Γ is a curve in V and Γ has contact of order μ_0 with V, then $\Gamma \subset V$.

Let C be a curve in V, and $\pi: X \longrightarrow C$ its desingularization, and $N \longrightarrow X$ the pull-back. By the lemma there is a formal extension to N, $\widetilde{F}$ of $f: N_0 \longrightarrow M$ which covers the map π. $\widetilde{F}$ is the formal pul-back of a map $\widetilde{\beta}: X \longrightarrow D$ such that $i \circ \widetilde{\beta} = \pi$. By Artin's Theorem we can find an analytic map $\beta: X \longrightarrow D$ so that β agrees with $\widetilde{\beta}$, and the pull-back F with $\widetilde{F}$, up to order μ_0. Thus, $i \circ \beta(N)$ has contact of order μ_0 with C, so $(i \circ \beta)(N) \subset V$. Since $i(D)$ intersects each branch of K in a curve, we must have $i(D) = K$.

REFERENCES

[1]. D. Burns, *"Some Background and Examples in Deformation Theory"*, in **Complex Manifold Techniques in Theoretical Physics,** Pitman, London (1979).

[2]. E. Horikawa, *"On Deformation of Holomorphic Maps I"*, **J. Math. Soc. Japan,** 25, 3, (1973), p. 372-396.

[3]. M. Kalka, *"Deformations of Submanifolds of Strongly-Negatively Curved Manifolds"*, **Math. Ann.** 251 (1980), p. 243-248.

[4]. Y. T. Siu, *"Complex Analyticity of Harmonic Maps, Vanishing and Lefschetz Theorems"*, in **J. Diff. Geom.** 25 (1982), p. 55-138.

[5]. J. Wavrick, *"Deformations and Analytic Equations"*, in **Proceedings of Symposia in Pure Mathematics,** Vol. 30, A.M.S. Providence (1977), p. 297-301.

Department of Mathematics
Tulane University
New Orleans, Louisiana 7 01 18
U.S.A.

A MARKOE

Computerized tomography and complex analysis

Acknowledgements

This paper is an expanded version of a talk given at the Workshop in Several Complex Variables which was part of the Third Colloquium in Mathematics sponsored by the Department of Mathematics, Centro de Investigacion y Estudios Avanzados del IPN, Mexico City.

I would like to thank Dr. E. Ramírez de Arellano and Prof. D. Sundararaman for their hospitality to me during the colloquium.

I would like to thank Dr. G. T. Herman for providing the material in Figures 1-7. My thanks also go to his co-authors and publishers, cited in the following papers, for permission to reprint these figures.

Figure 1 appeared as Figure 5 in L. Axel and G. T. Herman, *"Computerized Tomography and Nuclear Magnetic Resonance Imaging"*, **Abacus** 1 (1984) 30-41. Figures 2 and 3 appeared as Figures 2 and 4 in G. T. Herman, *"X-Ray Computed Tomography – Basic Principles"*, in *Three Dimensional Imaging Methods in Medicine and Biology*, (R. A. Robb, ed.), **CRC Press**, to appear. Figures 4-7 appeared as Figures 1, 3, 4, 5, respectively, in H. I. Goldberg, L. A. Bruno, G. T. Herman, C. R. Meyer and F. González-Scarano, *"Three-Dimensional CT Spine Reconstruction Exclusively Identifying Causative Lesion for Cervical Spondylotic Radiculopathy"*, **Journal of Neurosurgery**, to appear.

Contents

1. **Introduction**
2. **Photon Statistics**
3. **Transmission Tomography, CAT Scanners, and The Radon Transform**
4. **Single Photon Emission Tomography, The Attenuated Radon Transform and The Exponential Radon Transform**
5. **Analytic Continuation and Inversion of The Exponential Radon Transform**

1. Introduction

In computerized tomography projections of an object whose interior is unobservable are gathered at various angles. These projections constitute the Radon Transform, since they are integrals along lines of the function which describes the interior of the object. The tomographic process is completed by a computer algorithm which approximately inverts the Radon Transform. The best known tomographic process, CAT scanning, has revolutionized diagnostic medicine since its introduction about a decade and a half ago (the 1979 Nobel Prize in Medicine was shared by A. M. Cormack and G. N. Hounsfield for their work in developing tomographic scanners).

The term tomography originated in diagnostic medicine, but tomographic processes are used in Geology, Industrial Testing and other fields as well. At one time the volume of an iceberg was estimated using tomography! In computerized tomography (CT), x-rays are the most common mode of energy used for producing the projections needed to invert the Radon transform. But sound and magnetic fields are also used in ultrasound and NMR tomography, respectively. There is another form of tomography in which the energy source is located in the interior of the object rather than outside the object. This type of tomography is called Emission Tomography. The main mathematical result presented in this paper is related to Single Photon Emission Tomography which is modeled by a transform called the *attenuated Radon transform.* However, a good deal of attention will be paid to Transmission Tomography (CT) because of its intrinsic interest and because it provides an important analogy for the emission case.

Figure 1 presents an engineeering drawing of a typical CT scanner. The data necessary to produce the projections for the Radon Transform are gathered by the Data Acquisition/Detector Unit pictured in the gantry ring. The ring is then rotated to produce projections at various angles. Figure 2 was produced by a CT scanner, but not by tomographic reconstruction. It is medically equivalent to a standard skull x-ray and shows the usual super-position of features which obscure desired features in a standard x-ray. Figure 3 is a cross-section of the same skull as in Figure 2. But this cross-section was produced by tomographic reconstruction and clearly defines the features interior to the skull. Figure 4 is another classical x-ray, this time of a spinal column. Figure 5 shows a CT reconstruction of a cross-section of this spine. Several such cross-sections can be stacked together to form a three-dimensional picture of the interior. These three-dimensional pictures can be further processed to highlight individual features. This has been done to produce Figures 6 and 7 which show beautiful pictures of a portion of the spinal column. These pictures appear as if the spinal column had been surgically excised, but they actually were produced non-invasively by CT tomography.

The mathematical methods used in tomography are generally of a real analytic or harmonic analytic nature. But there is an approach to inverting the attenuated Radon Transform which depends on analytic continuation of holomorphic functions. Therefore, this subject seemed to be an appropriate choice to present at this colloquium.

We will always work in dimension 2, although the results have generalizations to dimension n. However, in dimension n $(n > 2)$ the appropriate transforms are x-ray transforms, not Radon transforms.

The references at the end will suggest some general works on tomography, as well as specialized papers related to the results presented here.

We start off by considering photon statistics in the next section. The following section is concerned with inverting the Radon transform. This is of interest because it shows the mathematical process behind CAT scanning and because it provides an important analogy for the inversion of the

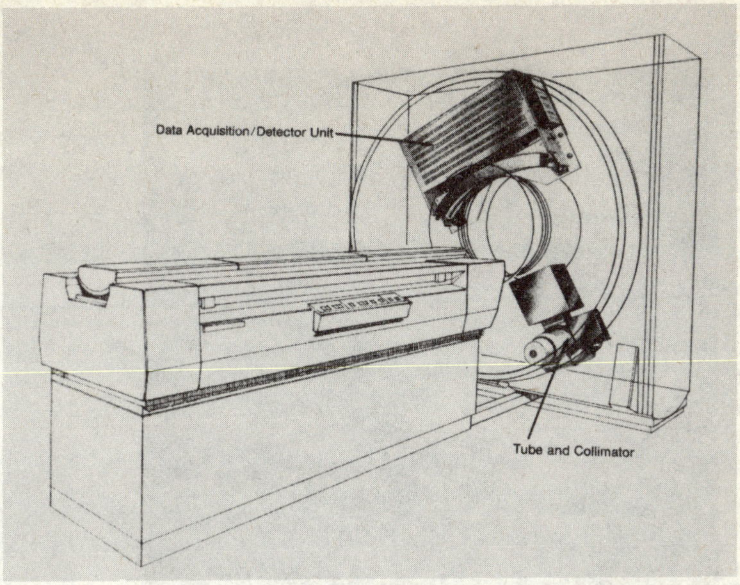

Figure 1: Engineeering drawing of typical CT scanner. The x-ray tube is shown at the lower right and the x-ray detector array and data acquisition system are shown at the upper left of the "gantry". The patient lies on the table; the table top can be slid into the reconstruction region in the middle of the gantry. In use, the x-ray tube and data acquisition system are rotated together around the patient to collect the set of projection data required for the reconstruction.
Reprinted with permission from 'Computerized Tomography and Nuclear Magnetic Resonance Imaging', by G.T. Hermann and L. Axel, Abacus 1984, Vol. 1 No. 2, pp. 30-41, Copyright Springer Verlag, New York.

attenuated Radon transform. The third section introduces Single Photon Emission Tomography and the associated attenuated Radon transform. Also the exponential Radon transform is introduced here because it is related to the constantly-attenuated Radon transform. A projection theorem due to F. Natterer relates the Fourier transform of the exponential Radon transform to the Fourier transform of the object on a portion of $\mathbb{C}^2$. This leads to a theorem on analytic continuation on $\mathbb{C}^2$ which is treated in the last section. A consequence of this continuation theorem is the invertibility of a class of attenuated Radon transforms.

2. Photon Statistics

We start with a brief foray into the photon statistics in order to derive the Radon transforms needed for the sequel.

A simple model of the behaviour of photon beams interacting with tissue is based on the intuitively appealing assumption that photons are absorbed or scattered in proportion to the density of the tissue through which they travel. Actually only mono-energetic photon beams obey this simple model, but good practical reconstructions based on this model exist. On the other hand some noise comes out in the reconstruction due to beam hardening and the spectrum of energies. If the photon beam originates outside the body to be diagnosed, then we can show that the photon projections yield the Radon transform. However, photon statistics will also be useful in emission tomography where the photons originate inside the body. A quantified version of our assumption is given by the Lambert-Beer Law:

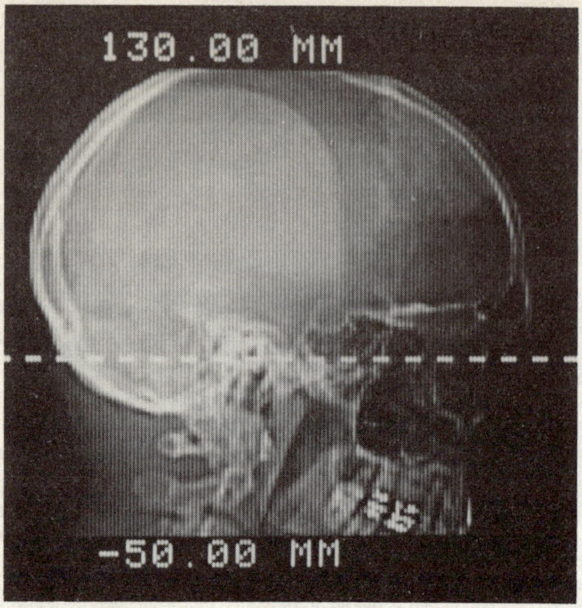

Figure 2. Scout View (a General Electric Trademark) obtained by sliding a patient through a stationary gantry of a CT scanner. A cross-section of interest is marked by a broken line.
Reprinted with permission from "X-Ray Computed Tomography—Basic Principles", to appear in the book 'Three Dimensional Imaging Methods in Medicine and Biology', R.A. Robb, ed. Copyright, CRC Press, Inc., Boca Raton. Fla.

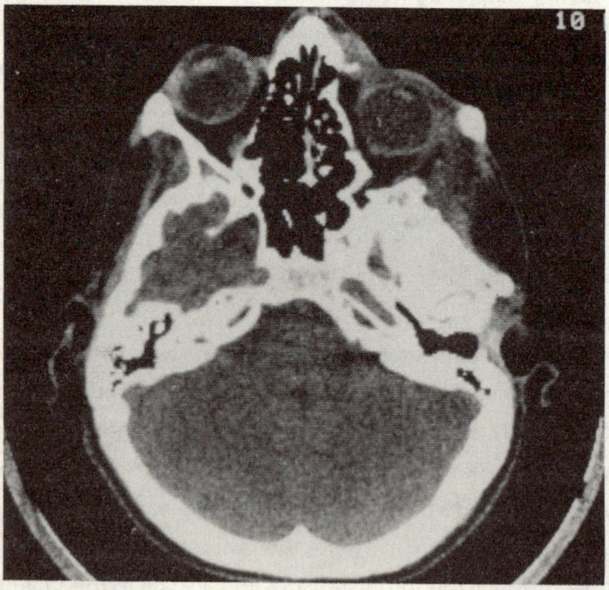

Figure 3. CT Reconstruction of the cross-section indicated in Figure 2.
Reprinted with permission from "X-Ray Computed Tomography—Basic Principles", to appear in the book 'Three Dimensional Imaging Methods in Medicine and Biology', R.A. Robb, ed., Copyright, CRC Press, Inc., Boca Raton, Fla.

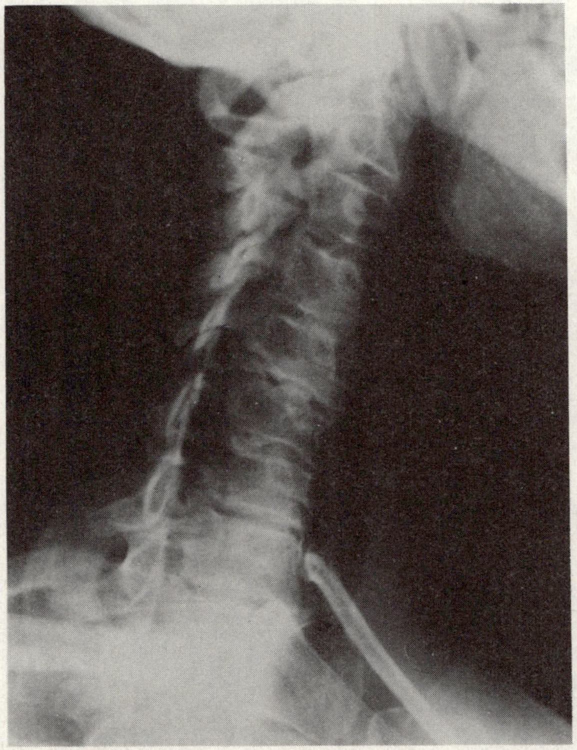

Figure 4. Cervical spine left posterior oblique projection. Large Lusckha joint osteophytes enroach into right C7-T1 neural foramen
Reprinted with permission from "Three Dimensional CT Spine Reconstruction Exclusively Identifying Caustic Lesion for Cervical Spondylotic Radiculopathy". H.I. Goldberg and F. González-Scarano, to appear. Journal of Neurosurgery.

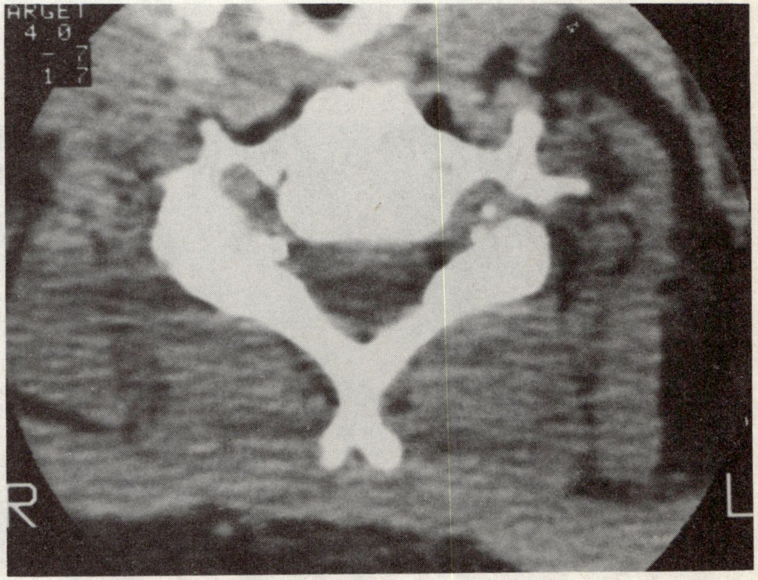

Figure 5. Transverse CT section at C6-7 neural foramen. Small nodular calcification present at both foramina.
Reprinted with permission from "Three Dimensional CT Spine Reconstruction Exclusively Identifying Caustic Lesion for Cervical Spondylotic Radiculopathy". H.I. Goldberg and F. González-Scarano, to appear. Journal of Neurosurgery.

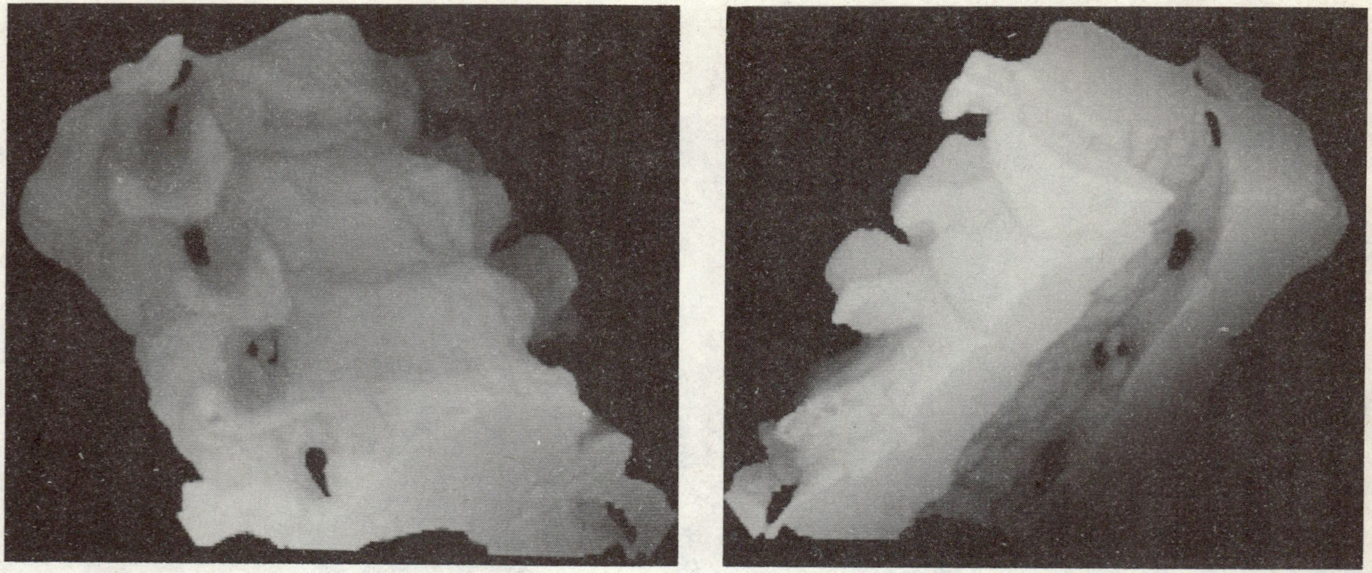

Figure 6. 3-D images of right neural foramina C4-T1. The lamina and spinous processes have been removed from the reconstructed images. Anterior oblique view (left) and posterior oblique view (right) demonstrate nodular bony bridge which crosses and constricts medial aspect of C6-7 foramen Luschka joint osteophytes present at C7-T1 which do not appear to cause any significant narrowing. The other neural foramina show no constriction.

Reprinted with permission from "Three Dimensional CT Reconstruction Exclusively Identifying Caustic Lesion for Cervical Spondylotic Radiculopathy". H.I. Goldberg and F. González-Scarano, to appear. Journal of Neurosurgery.

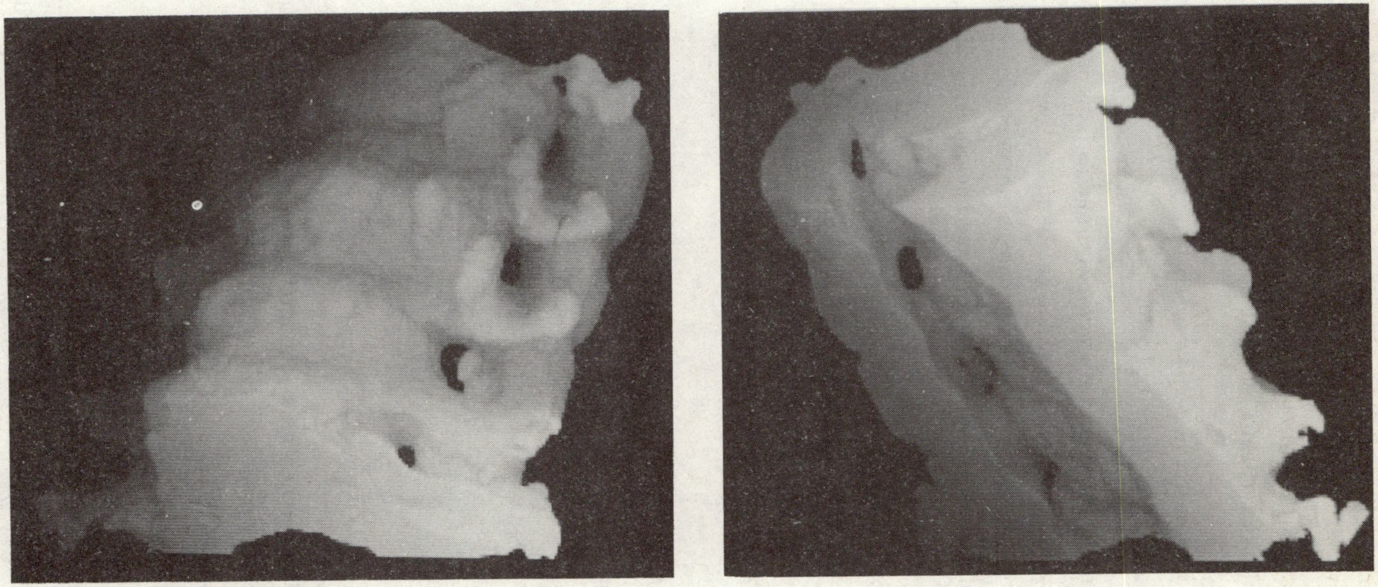

Figure 7. 3-D images of left neural foramina C4-T1. The lamina and spinous processes have been removed from the reconstructed images. Anterior oblique view (left) and posterior oblique view (right) demonstrate posterior bony excrescence projecting over outer aspect of C6-7 foramen. Luschka joint osteophytes impinge on inferior aspect of C7-T1 foramen. Other oblique views revealed the bony excrescence to be completely outside of the foraminal canal and that there was no significant narrowing of any of the foramina.

Reprinted with permission from 'Three Dimensional CT Spine Reconstruction Exclusively Identifying Caustic Lesion for Cervical Spondylotic Radiculopathy". H.I. Goldberg and F. González Scarano, to appear. Journal of Neurosurgery.

If photons travel along a straight line L, parametrized by distance s and if f(s) is the density of the medium at s on L, then the fraction of photons absorbed or scattered between s and s+ds is f(s) ds (accurate to first order).

To derive the connection between photon beams and line integrals, let E(L) be the event that a photon successfully traverses a linear segment L without being absorbed or scattered. Then, if L is short, we have to first order that

$$P(E(L)) = 1 - f(s)\,m(L)$$

Here P denotes probability, m denotes linear Lebesgue measure and s is the parameter of a point on L. Also we have used the complementary form of the Lambert-Beer law.

It is unlikely that the percentage of photons passing through disjoint segments should be stochastically dependent (at least from the point of view of classical mechanics). Thus, if L_1 and L_2 are disjoint segments, then

$$P(E(L_1 \cup L_2)) = P(E(L_1) \cap E(L_2)) = P(E(L_1))\,P(E(L_2)).$$

Now break up the path of the photon beam into small segments L_j and let ds_j denote $m(L_j)$. Let F be the probability that a photon which enters the tissue eventually exits without being absorbed or scattered. Then

$$F = P(E(L_1) \cap \ldots \cap P(E(L_k)) = \prod (1 - f(s_j') \, ds_j$$

Now use the fact that $1-x = \exp(-x)$ to first order to get

$$F = \exp(-\Sigma f(s_j)\, ds_j) \qquad \text{(to first order)}$$

Then take the limit as $ds \longrightarrow 0$ to get

$$F = \exp(-\int_L f(s)\, ds).$$

The probability F is approximately the fraction of photons which successfully traverse the tissue along L. Thus if $N_0(L)$ photons are emitted at the x-ray source and if $N_1(L)$ photons are detected after passing through the tissue, then

$$\int_L f(s)\, ds = -\ln(N_1(L)/N_0(L))$$

In other words, if one knows how many photons are emitted at the source in a certain direction and how many are detected after passing through the tissue, then one can estimate the line integral of the tissue density in that direction.

J. Radon [13] showed in 1917 how one can reconstruct f from all its line integrals. Unfortunately his algorithm involves differentiation and singular integration and is therefore not stable. In the next section we derive a stable inversion of this Radon transform (the Radon transform coincides with the x-ray transform, i.e., the k-plane transform for $k=1$, in dimension 2 but not in higher dimensions).

An interesting question which we do not pursue is the effect on the reconstruction of having only x-ray data from finitely many sources. A theorem of Smith, Solmon and Wagner [15] states that finitely many x-ray projections do not tell anything in the sense that given any fixed object and finitely many of its x-ray projections one can get another object with the same projections (from the finite number of directions) but which is very different from the original object.

On the other hand for low bandpass objects this result may be needlessly idealistic. For the pathological object must vary wildly near the boundary of the tissue in order to be different from the original object but still have many of the same projections. Therefore the pathological object will have most of its energy in high bandwidths and therefore the difference will probably be transparent after filtering, If the original object happened to have most of its energy in low bandwidths.

3. Transmission Tomography, CAT Scanners, and The Radon Transform

In CAT scanning, a beam of x-rays is passed through a transverse section of a patient. Some of the source photons are absorbed or scattered during transit. This loss of photons is described by the Lambert-Beer model which was discussed before. The surviving photons are collected in a collimated detector on the opposite side of the patient from the source. As we saw in Section 3:

$$\int_L f(s)\,ds = -\ln(N_1/N_0)$$

where f is the density function of the transverse section, N_0 is the number of photons emitted at the source and N_1 is the number detected. If this procedure is repeated at many angles, the resulting collection of line integrals forms an approximation to the Radon transform of f.

The Radon transform R is formally defined (on $L^1(\mathbb{R}^2)$) as follows:

Let $\theta \in S^1$ (the unit 1-sphere) and let s be in the line orthogonal to θ. Then,

$$Rf(s,\theta) = \int_{-\infty}^{\infty} f(s+t\theta)\,dt = \int_{L(S,\theta)} f(\sigma)\,d\sigma$$

where $L(S,\theta)$ = line through s with direction θ.

We will now give a method of inverting the Radon transform which does not depend on differentiation or singular integration as Radon's original inversion did.

Let FT denote the Fourier transform; if $f \in L^1(R^2)$, then

$$FT(f)(t) = \int_{R^2} \exp(-2\pi i x \cdot t) f(x)\, dx$$

Theorem. *(Projection Theorem of Helgason and Ludwig) The two-dimensional Fourier transform of* f *and the one-dimensional Fourier transform of* Rf *are related as follows:*
Let $\theta \in S^1$ *and let* s *be in the line orthogonal to* θ. *Then,*

$$FT(f)(s) = [FT(Rf)](s, \theta)$$

In other words, the Fourier transform of f restricted to the line orthogonal to θ, is proportional to the Fourier Transform of the Radon projection in the direction θ.

Remarks

1. If one generalizes to $\mathbb{R}^n$, then the transform defined here, integration over lines, becomes the x-ray transform and the projection theorem is still true. The Radon transform in higher dimensions is integration over hyperplanes. More generally, the k-plane transform is integration over k-planes. Then the x-ray transform is the 1-plane transform and the Radon transform is the n–1 plane transform.
2. The proof of the theorem is omitted. But it is really easy, just a simple application of change of variables in integration [Smith, Solmon, and Wagner 15, have a slightly more general result (page 1234 (3.1))].

It is now clear that the Radon transform is invertible. In practice one could approximate the density function f by taking Fourier transforms of the projection data thus obtaining an approximation to the Fourier transform of f. Then f could be recovered by inverse Fourier transformation.

Actual CAT scanners generally use a filtered convolution algorithm. This is mathematically similar to the method described here, but avoids actual Fourier transformation by doing convolutions in the "time" domain.

4. Single Photon Emission Tomography, The Attenuated Radon Transform

In single photon emission tomography a photon-emitting substance is injected into a patient. Then the emitted photons are collected by a detector outside the patient. If this is done at many angles, then an approximation to the attenuated Radon transform is obtained.

Let f be the photon density of the injected material and let μ be the density function of the ambient tissue (contrast this to CAT scanning where f was the tissue density). The ambient tissue attenuates the emitted photon beams because of absorption and scatter. If we take a line L with direction $\theta \in S^1$ and consider a point $s + t\theta$ on L, then the Lambert-Beer law in photon statistics tells us that about

$$\exp(-\int_t^{\infty} \mu(s+\tau\theta)\, d\tau)$$

photons of the $f(s+t\theta)$ that started out at $s+t\theta$, will be detected in the detector with direction θ. But if we add up (i.e., integrate) along L, we get the total number of photons detected by the detector with direction θ to be

$$\int_{-\infty}^{\infty} f(s+t\theta)\exp\left(-\int_{t}^{\infty} \mu(s+\tau\theta)\,d\tau\right) dt \qquad (1)$$

We define the attenuated Radon transform $R_\mu f$ at (s, θ) to be the integral (1).

The problem of inverting the general attenuated x-ray transform is open. But if the attenuation is constant, then several people have shown that the inversion is possible [Tretiak and Metz 5, Bellini, Piacentini and Cafforio 1, Quinto 4, Markoe 2]. I will show you a way to do it that uses complex analysis.

Actually it suffices to show that a related transform is invertible. The related transform is called the exponential Radon transform [Tretiak and Metz, 5] and is defined by

$$Q_\mu f(s, \theta) = \int_{-\infty}^{\infty} f(s+t\theta)\exp(\mu t)\,dt$$

where μ is a constant (the constant attenuation, if we are thinking about that situation).

In my paper [2] I show that the exponential Radon transform is invertible. Other people [Tretiak and Metz 5, Natterer 3] had observed that this is all that is really necessary to invert the constantly attenuated Radon transform (in fact $R_\mu f(s, \theta) = \exp(-\mu(1-|s|^2)^{1/2})\, Q_\mu f(s, \theta)$).

But I also showed that a little trick proves that the attenuated Radon transform is invertible provided that the convex hull of the emitting photons is contained in a region of constant attenuation. This is an improvement over the constant attenuation case, but of course is not the whole story. In any case I refer you to [2] for the details of this trick.

The analogue for exponential Radon transforms to the projection theorem was discovered by F. Natterer:

Theorem. *(Natterer's Attenuated Projection Theorem)* [3]. *If* $\theta \in S^1$ *and* $\zeta \perp \theta$, *then*

$$FT(\zeta + i\mu\theta) = FT(Q_\mu F)(\zeta, \theta).$$

If $Q_\mu f$ *is zero, then* $FT(f)$ *is zero on the subset of* $\mathbb{C}^2$:

$$A^\mu = \left\{\zeta + i\mu\theta : \theta \in S^1 \text{ and } \zeta \perp \theta\right\}.$$

In practice f has compact support, and this restriction is assumed from here on. Then $FT(f)$ is an entire holomorphic function by the easy part of the Paley-Wiener theorem. So we are in the domain of complex analysis. In the next section we will see how A^μ is a set of uniqueness for holomorphic functions in $\mathbb{C}^2$. This will prove that the exponential Radon transform is invertible.

5. Analytic Continuation and Inversion of The Exponential Radon Transform

The specific extension problem considered here is to continue a holomorphic function from the set A^μ defined previously to $\mathbb{R}^2$. A consequence of this extension will be that A^μ is set of uniqueness for entire holomorphic functions. Only a sketch of the process is given; full details may be found in [2].

The idea in the analytic continuation is to define a family of "almost" analytic discs parametrized by $\mathbb{R}^2$ whose center is the parameter and whose boundary is contained in A^μ. These discs are only "almost" analytic because the transformation will involve a square root. However, this problem is overcome by decomposing the function to be continued into even and odd parts. Then there will be enough analyticity to use the Cauchy integral formula to continue the function from A^μ to $\mathbb{R}^2$.

Once the function is continued from A^μ to $\mathbb{R}^2$, it is trivial to show that A^μ is a set of uniqueness for entire functions in the sense that if two entire functions agree on A^μ, then they agree on $\mathbb{C}^2$. In fact it is enough to show that if F is zero on A^μ, then F is zero on $\mathbb{C}^2$. But the analytic continuation will show that any entire F zero on A^μ also vanishes on $\mathbb{R}^2$; finally the Cauchy-Riemann equations give the global uniqueness.

To begin the construction, define a family of plane curves parametrized by $s \in \mathbb{R}^2$:

$$\theta \longrightarrow \Gamma_s(\theta) = [s_1^2 + s_2^2 + \mu^2]^{1/2} \sin\theta + i\mu\cos\theta$$

Then embed these curves in $\mathbb{C}^2$ using the transformation $Z{:}\mathbb{C}\times\mathbb{R}^2 \longrightarrow \mathbb{C}^2$:

$$Z_1(w, s) = w$$

$$Z_2(w, s) = [s_1^2 + s_2^2 - w^2]^{1/2}$$

The branch of the square root is arbitrary, except that if $w \in \Gamma_s$, then

$$Z(w, s) = [s_1^2 + s_2^2 + \mu^2]^{1/2} \cos\theta + i\mu\sin\theta$$

An easy calculation shows that the images of the curves are contained in A^μ and their centers transform to the parameter s in $\mathbb{R}^2$. Then if we take an even holomorphic function G in $\mathbb{C}^2$, we can pull back analytically to $\mathbb{C}^2$ via the transformation Z and use Cauchy's integral formula on the ellipse Γ_s to get the continuation of G to s. Full details of this construction can be found in [2] along with an explicit integral formula for constructing the continuation. The injectivity of the exponential Radon transform is established by this continuation theorem in conjunction with Natterer's projection theorem. A method of inverting the exponential Radon transform is to use Natterer's projection theorem on the Fourier transform of $Q_\mu f$ to obtain the integrand of the integral formula referred to above. The result of the integral formula gives the Fourier transform of f on $\mathbb{R}^2$. Then Fourier inversion can be used to recover f.

SPECIALIZED REFERENCES

[1]. S. Bellini, M. Piacentini and C. Cafforio, *Compensation of Tissue Absorption in Emission Tomography.* **IEEE Trans. Acoustics, Speech and Signal Processing,** ASSP-27 (1979) 213-118.

[2] A. Markoe, *Fourier Inversion of The Attenuated X-Ray Transform,* **Siam J. Math. Anal.,** to appear.

[3]. F. Natterer, *On the Inversion of the Attenuated Radon Transform,* **Numer. Math.** 32 (1979) 432-438.

[4]. E. T. Quinto, *The Invertibility of Rotation Invariant Radon Transforms,* **J. Math. Anal. Appl.** 51 (1983) 510-522.

[5]. O.J. Tretiak and C. Metz, *The Exponential Radon Transform,* **Siam J. Appl. Math.** 39 (1980) 341-354.

GENERAL REFERENCES

[6]. R. A. Brooks and G. Dichiro, *Principles of Computer-Assisted Tomography (CAT) in Radiographic and Radio-Isotopic Imaging,* **Phys. Med. Biol.** 21 (1976) 689-732.

[7]. A. M. Cormack, *Representation of A Function by Its Line Integrals, with Some Radiological Applications,* **J. Appl. Phys.** 34 (1963) 2722-2727.

[8]. S. R. Deans, *The Radon Transform and Some of Its Applications,* John Wiley and Sons, New York, (1983).

[9]. S. Helgason, *The Radon Transform,* Birkhauser, Boston, Basel, Stuttgart, (1980).

[10]. G. T. Herman, *Image Reconstruction from Projections: The Fundamentals of Computerized Tomography,* Academic Press, New York, (1980).

[11]. G. N. Hounsfield, *Computerized Transverse Scanning (Tomography): Part I. Description of System,* **Brit. J. of Radiology** 46, (1973), 1016-1022.

[12]. R. M. Merserau and A. V. Oppenheim, *Digital Reconstruction of Multidimensional Signals from Their Projections,* **Proc. IEEE** 62 (1974) 1319-1338.

[13]. J. Radon, *Über die Bestimmung von Funktionen durch ihre Integralwerte längs Gewisser Mannigfaltigkeiten. Berichte Sächsiche Akademie der Wissenschaften.* Leipzig, **Math-Phys.** Kl., 69 (1917) 262-267. [This paper may be found in Reference 9 and in translation in Reference 8].

[14]. H. J. Scudder, *Introduction to Computer-Aided Tomography,* **Proc. IEEE** 66 (1978) 628-637.

[15]. K. T. Smith, D. C. Solmon and S. L. Wagner, *Practical and Mathematical Aspects of the Problem of Reconstructing Objects from Radiographs,* **Bull. Am. Math. Soc.** 83 (1977) 1227-1270.

Department of Mathematics and Physics
Rider College
Lawrenceville, New Jersey 08468,
USA

A NAGEL

Non-isotropic metrics on boundaries of domains of finite type

This is a summary of joint work with E. M. Stein and S. Wainger on non-isotropic metrics and pseudometrics on the boundaries of domains of finite type $D \subset \mathbb{C}^{n+1}$. Our work has been directed towards generalizing two kinds of results from the better understood case of strictly pseudoconvex domains. The first deals with boundary behavior of functions in $H^p(D)$, $0 < p \leqslant \infty$, and centers around the notion of approach region, and associated maximal functions and area integrals. The second involves obtaining estimates for certain integral operators on the boundary ∂D of D, such as parametricies for $\Box_b$. In this note I shall try to show how both the geometry of the approach regions, and the *singularities* of the integral kernels can be described in *terms* of a family of balls on the boundary of D defined by a non-isotropic metric or pseudometric.

§1: A Strictly Pseudoconvex Example

In order to motivate our work, I will begin by recalling some known results in the case of the standard model for strictly pseudoconvex domains, the Siegel upper half-space. Thus we let:

$$U = \{z = (z_1, \ldots, z_{n+1}) \in \mathbb{C}^{n+1} \mid \operatorname{Im} z_{n+1} > \sum_{j=1}^{n} |z_j|^2\}.$$

Then U is a strictly pseudoconvex domain which is biholomorphically equivalent via a Caley transformation to the unit ball B in $\mathbb{C}^{n+1}$.

We begin with questions relating to the boundary behavior of holomorphic functions on U. For $0 < p < \infty$, let $H^p(U)$ denote the space of holomorphic functions F on U for which

$$\sup_{r>0} \int_{\mathbb{P}^n \times \mathbb{R}} |F(z_1, \ldots, z_n, t + i(r + \sum_{j=1}^{n} |z_j|^2))|^p \, dt\, dz = \|F\|^p_{H^p} < \infty.$$

Since every $F \in H^p(U)$ is harmonic when viewed as a function on a smoothly bounded domain in $\mathbb{R}^{2n+2} = \mathbb{C}^{n+1}$, one can apply to F results of standard potential theory. In particular, it follows that every such F has non-tangential limits at almost every point of the boundary of U. Here, non-tangential approach to a point $\rho \in \partial U$ means approach through a region

$$\Gamma_\alpha(\rho) = \{z \in U \mid |z - \rho| < (1+\alpha)\, h(\rho)\}$$

where $\alpha > 0$ is the "aperture", and

$$h(z) = \operatorname{Im} z_{n+1} - \sum_{j=1}^{n} |z_j|^2$$

is the "height" of a point $z \in U$ above the boundary. Geometrically, the region $\Gamma_\alpha(\rho)$ is essentially a non-tangential cone with vertex at ρ.

By now, however, it is well known that this classical result can be improved since F is actually holomorphic, and that non-isotropic phenomena appear. In 1969 Koranyi [2] introduced the notion of admissible approach regions for certain symmetric domains including U, and this was extended in 1972 by E. M. Stein [6] to general smoothly bounded domains in $\mathbb{C}^n$. For our particular example, these approach regions are defined as follows.

Let $\rho \in \partial U$ and let $\alpha > 0$. Set:

$$\mathcal{a}_\alpha(\rho) = \{z \in U \mid |\langle z-\rho, U_\rho\rangle| < (1+\alpha)h(z),\ |z-\rho|^2 < \alpha h(z)\}.$$

where U_ρ is the exterior unit normal at ρ, and $\langle\ ,\ \rangle$ is the usual Hermitian inner product on $\mathbb{C}^{n+1}$.

The condition $|\langle z-\rho, U_\rho\rangle| < (1+\alpha)h(z)$ is clearly a condition only on the projection of z onto the complex line containing U_ρ (and iU_ρ), and it simply says that this projection must be contained in a non-tangential cone. The condition $|z-\rho|^2 < \alpha h(z)$ allows second-order tangential approach in all directions. Thus the region $\mathcal{a}_\alpha(\rho)$ allows approach to the boundary point ρ which is non-tangential in the "real" direction iU_ρ, but allows second-order tangential approach in the n complex orthogonal directions.

Using the regions $\mathcal{a}_\alpha(\rho)$, one can now prove a strengthening of the classical result of Fatou on the existence of limits, as well as a generalization of the more quantitative result of Hardy and Littlewood on the non-tangential maximal function. Thus, for F holomorphic in U, $\rho \in \partial U$, and $\alpha > 0$, define

$$\mathfrak{M}_\alpha F(\rho) = \sup_{z \in \mathcal{a}_\alpha(\rho)} |F(z)|$$

Also, we say F has admissible limit at ρ if for every $\alpha > 0$,

$$\lim_{\substack{z \to \rho \\ z \in \mathcal{a}_\alpha(\rho)}} F(z) \qquad \text{exists.}$$

Theorem A:

a) Suppose $0 < p < \infty$ *and* $\alpha > 0$*. Then there is a constant* $A_{p,\alpha} < \infty$ *so that if* $F \in H^p(U)$,

$$\|\mathfrak{M}_\alpha F\|_{L^p(\partial U)} \leqslant A_{p,\alpha} \|F\|_{H^p(U)}$$

b) If $0 < p < \infty$ *and* $F \in H^p(U)$ *then* F *has admissible limits at almost every* $\rho \in \partial U$.

For a proof of this theorem, as well as results on appropriate variants of the Lusin area integral, see Stein [6].

We next turn to questions relating to integral operators on ∂U. In $\mathbb{R}^m$, $m \geqslant 3$, it is well known that a fundamental solution for the Laplace operator Δ is given by

$$N(x) = c_m |x|^{-m+2}, \qquad c_n = \Gamma(\tfrac{n}{2})/2(2-n)\pi^{n/2}.$$

Thus, if $f \in C_0^\infty(\mathbb{R}^m)$ and if

$$K(x, y) = c_m |x-y|^{-m+2}$$

then

$$u(x) = Tf(x) = \int_{\mathbb{R}^m} k(x, y) f(y) dy$$

satisfies $\Delta u = f$.

We note for future reference that we have the estimate

$$|K(x, y)| = c_m \frac{d(x, y)^2}{|B(x, d(x, y))|}$$

where $d(x, y) = |x-y|$ is the Euclidian distance from x to y, $B(x, \delta)$ is the Euclidian ball centered at x of radius δ, and $|E|$ denotes the volume of a set $E \subset \mathbb{R}^m$.

There is an analogue of the Laplace operator on ∂U which reflects the complex nature of the situation. This operator, the Kohn Laplacian $\Box_b$, acts on differential forms on ∂U, and is given by

$$\Box_b = \bar\partial_b \bar\partial_b^* + \bar\partial_b^* \bar\partial_b$$

where $\bar\partial_b$ is the usual $\bar\partial$ operator on $\mathbb{C}^{n+1}$ restricted to the boundary, and $\bar\partial_b^*$ is the formal adjoint. When restricted to q forms, $\Box_b^{(q)}$ acts diagonally, and is a second-order differential operator given as follows. We can parametrize ∂U by $\mathbb{C}^n \times \mathbb{R}$ via the map

$$(z_1, \ldots, z_n, t) \longrightarrow (z_1, \ldots, z_n, t+i \sum_{j=1}^n |z_j|^2)$$

so that differential operators on $\mathbb{C}^n \times \mathbb{R}$ correspond to differential operators on ∂U. On $\mathbb{C}^n \times \mathbb{R}$ consider the operators

$$X_j = \frac{\partial}{\partial x_j} + \partial y_j \frac{\partial}{\partial t}, \qquad Y_j = \frac{\partial}{\partial y_j} - 2x_j \frac{\partial}{\partial t} \qquad 1 \leqslant j \leqslant n$$

and $T = \frac{\partial}{\partial t}$. Set

$$Z_j = \frac{1}{2}(x_j - iy_j), \qquad \bar Z_j = \frac{1}{2}(x_j + iy_j) \qquad 1 \leqslant j \leqslant n$$

and define

$$\mathcal{L}_\alpha = -\frac{1}{2}\sum_{j=1}^{n}(z_j\bar{z}_j + \bar{z}_j z_j) + i\alpha T.$$

Then
$$\square_b^{(q)} = \mathcal{L}_\alpha$$

where $\alpha = n-2q$.

Except for certain exceptional values of α, there is a fundamental solution for the operator $\mathcal{L}_\alpha$. Thus, for $\alpha \neq \pm n, \pm(n+2), \pm(n+4), \ldots,$ set

$$\psi_\alpha(\rho, t) = C_\alpha(|\rho|^2 - it)^{-\frac{(n+\alpha)}{2}}(|\rho|^2+it)^{-\frac{(n-\alpha)}{2}}$$

where $C_\alpha = \Gamma(\frac{n+\alpha}{2})\Gamma(\frac{n-\alpha}{2})/2^{2-n}\pi^{n+1}$. Then Folland and Stein [1] proved

Theorem B: *If* $\alpha \neq \pm n, \pm(n+1), \ldots,$ ψ_α *is a fundamental solution for* $\mathcal{L}_\alpha$. *If* $f \in C_0^\infty(\mathbb{C}^n \times \mathbb{R})$, *if*

$$K((z, t); (\omega, s)) = \psi_\alpha(z-\omega, t-s-2\,\mathrm{Im}\langle z, \omega\rangle),$$

and if

$$u(z, t) = Tf(z, t) = \int_{\mathbb{C}^n \times \mathbb{R}} f(\omega, s)\, K((z, t); (\omega, s))\, d\omega\, ds$$

then $\mathcal{L}_\alpha u = f$.

I now want to describe the connection between the kinds of results typified by Theorems A and B, and the geometry of a family of balls on ∂U. Again, we shall identify ∂U with $\mathbb{C}^n \times \mathbb{R}$. Then, if $\rho_1 = (z, t)$ and $\rho_2 = (\omega, s)$ are two points of ∂U, set

$$\rho(\rho_1, \rho_2) = \sup\{|z-\omega|, \ |t-s-2\,\mathrm{Im}\langle z, \omega\rangle|^{1/2}\}$$

It is easy to check that ρ is a pseudometric, i.e., there is a constant $C < \infty$ so that if $\rho_1, \rho_2, \rho_3 \in \partial U$,

(i) $\rho(\rho_1, \rho_2) \geqslant 0$ and $\rho(\rho_1, \rho_2) = 0$ if and only if $\rho_1 = \rho_2$.

(ii) $\rho(\rho_1, \rho_2) = \rho(\rho_2, \rho_1)$.

(iii) $\rho(\rho_1, \rho_3) \leqslant C[\rho(\rho_1, \rho_2) + \rho(\rho_1, \rho_3)]$.

Moreover, if we set

$$B(\rho, \delta) = \{\xi \in \partial U \,|\, \rho(\rho, \xi) < \delta\}$$

then $B(\rho, \delta)$ is a "ball" centered at ρ which is essentially a parallelepiped centered at ρ with side $2\delta^2$ in the real direction iU_ρ, and sides 2δ in the complementary orthogonal complex directions.

The first observation to make is that the admissible approach region $\mathcal{A}_\alpha(\rho)$ can be described in terms of the family of balls $\{B(\rho, \delta)\}$, the height function $h(z)$, and the normal projection $\pi: U \longrightarrow \partial U$ given by

$$\pi(z_1 \ldots z_{n+1}) = (z_1 \ldots z_n, \operatorname{Re} z_{n+1} + i \sum_{j=1}^{n} |z_j|^2) = (z_1, \ldots, z_n, z_{n+1} - ih(z))$$

Thus if we define

$$\widetilde{\mathcal{A}}_\alpha(\rho) = \{z \in U \,|\, \pi(z) \in B(\rho, \alpha h(z)^{1/2})\}$$

then there are constants $C_1, C_2 > 0$ so that for $\rho \in \partial U$ and $\alpha > 0$

$$\mathcal{A}_{C_1, \alpha}(\rho) \subset \widetilde{\mathcal{A}}_\alpha(\rho) \subset \mathcal{A}_{C_2 \alpha}(\rho).$$

Thus the approach regions $\mathcal{A}_\alpha(\rho)$, and also the admissible maximal functions $\mathfrak{M}_\alpha$ are determined by the family of balls $\{B(\rho, \delta)\}$.

Next, the proof of Theorem A can proceed by introducing a Hardy-Littlewood-type maximal function on the boundary of U. For $g \in L^1_{loc}(\partial U)$ set:

$$Mg(\rho) = \sup_{\delta > 0} \frac{1}{|B(\rho, \delta)|} \int_{B(\rho, \delta)} |g(\omega, s)| \, d\omega \, ds$$

Stein's idea in [5] is to dominate the admissible maximal function $\mathfrak{M}_\alpha F(\rho)$ by the composition of the operator M with the ordinary non-tangential (or even the radial) maximal function of F. Since the last operator is known to be bounded, one wants to show that M is of weak type (1, 1) and bounded on L^p if $1 < p \leqslant \infty$. This can be done with the usual proof of the Hardy-Littlewood maximal function theorem, *if* the family of balls $\{B(\rho, \delta)\}$ satisfies two crucial properties:

$$\left.\begin{array}{l} \text{(a) if } B(\rho_1, \delta) \cap B(\rho_2, \delta) \neq \emptyset \text{ then } B(\rho_1, \delta) \subset B(\rho_2, C\delta) \\ \text{(b) } |B(\rho, 2\delta)| \leqslant C\,|B(\rho, \delta)|. \end{array}\right\} (*)$$

Here C is a constant independent of ρ_1, ρ_2, and δ.

Thus, at least in part, the proof of Theorem A is reduced to certain geometric questions about the family of balls $\{B(\rho, \delta)\}$. In this particular example, (a) and (b) are easily verified, since the balls are defined by a pseudometric (which gives (a)), and one computes that

$$|B(\rho, \delta)| \approx \delta^{2n+2}$$

(which gives (b)).

Turning to the fundamental solution of $\square_b$ given in Theorem B, it is clear that we have the estimate

$$|K((z,t),(\omega,s))| \leqslant C_\alpha \frac{\rho(\rho_1,\rho_2)^2}{|B(\rho,\rho(\rho_1,\rho_2))|}$$

where $\rho_1 = (z, t)$, $\rho_2 = (\omega, s)$. This is exactly analogous to the estimates for the Newtonian potential which is the fundamental solution of the Laplace operator. Once one has this estimate, and analogous estimates in derivatives of K, one can proceed in a standard manner to prove that the operator T of Theorem B is actually a continuous operator between appropriate function spaces. For details see the paper of Folland and Stein [1].

§2. Domains of Finite Type

To what extent can the theory sketched above be generalized to more general domains in $\mathbb{C}^{n+1}$? For *any* smooth domain $D \subset \mathbb{C}^{n+1}$ one can define a family of balls on the boundary, $\{B(x,\delta)\}$, which are of size δ^2 in the real direction and size δ in the complex directions, so that the analogue of Theorem A holds. This was done by Stein in [5]. However, these results are not optimal if the domain is not strictly pseudoconvex. This is suggested by the following simple example.

Let

$$D = \{(z_1, z_2) \in \mathbb{C}^2 \mid \operatorname{Im} z_2 > |z_1|^{2k}\}$$

where $k > 1$. This domain is strictly pseudoconvex at all boundary points where $z_1 \neq 0$, but is only weakly pseudoconvex on the line when $z_1 = 0$. Suppose F is a bounded holomorphic function on D and suppose

$$\lim_{y \to 0^+} F(0, iy) \quad \text{exists.}$$

Then, applying the one-variable Schwarz lemma to the function

$$G(\rho) = F(\rho, iy) - F(0, iy)$$

one easily sees that F has a limit at $(0, 0)$ along curves making any order of contact less than $2h$ in the complex z_1 direction. This suggests that the optimal approach region at $(0, 0)$ should have $2k$-order contact with ∂D, rather than just the second order. Also if we expect that the approach regions can be defined in terms of a family of balls on ∂D, this suggests that the balls centered at $(0, 0)$ should be of size δ^{2k} in the real direction, and δ in the complex directions. On the other hand, when $z_1 \neq 0$, the domain is strongly pseudoconvex, and the balls on the boundary should be essentially of size δ^2 and δ, at least for small δ.

This raises the following question. Can one find a smoothly-varying family of balls on the boundary of D which reflects the degeneracy of the Levi form in the sense that for $z_1 \neq 0$ and small δ the balls are of size δ^2 and δ, while along the line of degeneracy $z_1 = 1$, the balls are of size δ^{2k} and δ? One would want such a family of balls to have the property that if new approach regions are defined using the balls, then the analogue of Theorem A should be true. In light of Theorem B, one would also like to show that the integral kernel of a fundamental solution or at least a parametrix for $\Box_b$ on D can be estimated in terms of this new family of balls. It turns out that one can define such a family of balls, and in fact this can be done on the boundary of any domain $D \subset \mathbb{C}^{n+1}$ of finite type. In order to see how to make this definition, I first show how to construct metrics out of certain families of vector fields on $\mathbb{R}^N$, and then I show how to apply their construction in the case of domains in $\mathbb{C}^n$.

Let $\Omega \subset \mathbb{R}^N$ be a connected open set, and let $X_1, \ldots, X_p$ be real C^∞-vector fields on $\overline{\Omega}$. Let $X^{(1)} = \{X_1, \ldots, X_n\}$ be the collection of these vector fields, let $X^{(2)} = \{\ldots [X_1, X_2], \ldots\}$ be the collection of the commutators of the vector fields, and in general, let $X^{(k)} = \{\ldots, [X_1, [X_i, \ldots [X_{i_{k-1}}, X_{i_k}] \ldots]], \ldots\}$ be the collection of commutators of length k. We shall suppose that the given vector fields satisfy a Hörmander condition of *finite type* m. This means that at every point of $\overline{\Omega}$, the elements of $X^{(1)}$, $X^{(2)}, \ldots, X^{(m)}$ span the tangent space at x.

We now define a metric ρ on Ω as follows. If $x, y \in \Omega$, we say $\rho(x, y) < \delta$ if and only if there is an absolutely continuous map $\varphi\colon [0, 1] \longrightarrow \Omega$, with $\varphi(0) = x$, $\varphi(1) = y$,

$$\varphi'(t) = \sum_{j=1}^{p} a_j(t)\, X_j(\varphi(t))$$

almost everywhere, with $|a_j(t)| < \delta$, $0 \leqslant t \leqslant 1$, $1 \leqslant j \leqslant p$.

It is a fact first found by Carathéodory that since the vector fields are of finite type, any two points of Ω *can* be joined by a curve φ with φ' in the span of $X_1, \ldots, X_p$. Then ρ is a metric on Ω. We set

$$B(x, \delta) = \{y \in \Omega \,|\, \rho(x, y) < \delta\}.$$

Then it is clear that

$$B(X_1, \delta) \cap B(X_2, \delta) \neq \emptyset \Longrightarrow B(X_1, \delta) \subset B(X_2, 3\delta)$$

so property (a) of (*) is satisfied.

Now let $Y_1, \ldots, Y_q$ be some ennumeration of the vector fields belonging to $X^{(1)}, \ldots, X^{(m)}$. If Y_j is an element of $X^{(m)}$, we say the degree of Y_j, d_j, is k. If $I = (i_1, \ldots, i_N)$ is any N-tuple of integers, with $1 \leqslant i_j \leqslant q$, we let

$$d(I) = d_{i_1} + d_{i_2} + \ldots + d_{i_N}$$

and we define

$$\lambda_I(x = \det(Y_{i_1}, \ldots, Y_{i_N})(x)$$

where if $Y_j = \sum_{k=1}^{N} a_{jk}(x) \frac{\partial}{\partial x_k}$,

$$\det(Y_1, \dots, Y_N)(x) = \{\det a_{jk}(x)\}.$$

We can then prove

Theorem 1: *There are constants* C_1 *and* $C_2 > 0$ *so that for all* $x \in \Omega$ *and all small* $\delta > 0$

$$C_1|B(x, \delta)| \leq \sum_I |\lambda_I(x)| \delta^{d(I)} \leq C_2|B(x, \delta)|.$$

then the sum is taken on all N-*tuples* I.

As a corollary, we see that

$$|B(x, 2\delta)| \leq 2^D \frac{C_2}{C_1} |B(x, \delta)|,$$

where $D = \sup_I d(I)$. Thus the doubling property (b) of (*) is satisfied.

We now apply the theory to the case of a domain $D \subset \mathbb{C}^{n+1}$ with smooth boundary of finite type. We shall work locally near a fixed point p at ∂D, which we can view as an open set in $\mathbb{R}^{2n+1}$. Let $L_1, \dots, L_n$ be linearly independent tangential holomorphic vector fields near p. Write

$$L_j = X_j + iX_{n+j} \qquad 1 \leq j \leq n$$

where $X_1, \dots, X_{2n}$ are real. To say that D is of finite type m is exactly the same as saying that the vector fields $X_1, \dots, X_{2n}$ satisfy a Hörmander condition of type m. Thus we can apply our theory, and construct a metric, and an associated family of balls on ∂D.

We want to give an alternate description of these balls. Locally near p we can choose a smooth vector field T so that $X_1, \dots, X_{2n}$, and T form a basis for the tangent space at every point ρ. Then every h*th*-order commutator $[X_1, [X_{i_1}, \dots [X_{i_{h-1}}, X_{i_h}] \dots]]$ can be written uniquely as a linear combination of $X_1, \dots, X_{2n}$, and T. Let $\lambda_{i_1 \dots i_n}(\rho)$ be the coefficient of T, and set

$$\Lambda_j(\rho) = \sum_{b \leq j} |\lambda_{i_1 \dots i_n}(\rho)|$$

$$\Lambda(\rho, \delta) = \sum_{j=2}^{m} \Lambda_j(\rho) \delta^j$$

Note that if ρ is a strictly pseudoconvex boundary point, $\Lambda_2(\rho) \neq 0$, so for small δ, $\Lambda(\rho, \delta) \approx \delta^2$. If ρ is a point of type m, so that $\Lambda_2(\rho) = \dots = \Lambda_{m-1}(\rho) = 0$, $\Lambda_m(\rho) \neq 0$, then $\Lambda(\rho, \delta) \approx \delta^m$.

We can now define a family of balls in terms of the exponential map given by the vector fields $X_1, \dots, X_{2n}, T$.

Thus, for $\rho \in \partial D$, let

$$\widetilde{B}(\rho, \delta) = \{\eta \in \partial D \mid \eta = \exp[\sum_{j=1}^{2n} \alpha_j X_j + \gamma T](\rho) \text{ where } |\alpha_j| < \delta, \ |\gamma| < \Lambda(\rho, \delta)\}$$

Thus $\widetilde{B}(\rho, \delta)$ is the image under this exponential map of a box which is of size δ in the complex direction $X_1, \ldots, X_{2n}$, and of size $\Lambda(\rho, \delta)$ in the real direction T. Moreover, these new balls are essentially the same as the balls defined earlier. We have

Theorem 2: *If* $\Sigma \subset \partial D$ *is compact, there are constants* C_1 *and* C_2 *so that if* $\rho \in \Sigma$

$$B(\rho, C_1\delta) \subset \widetilde{B}(\rho, \delta) \subset B(\rho, C_2\delta).$$

We now indicate how Theorem A generalizes to this situation. Let W be a sufficiently small tabular neighborhood of ∂D and let

$$\pi: W \cap D \longrightarrow \partial D$$

be a normal projection. For $z \in W \cap D$, let $h(z)$ be the distance of z from ∂D, and let $D(z)$ be the unique solution of the equation

$$\Lambda(\pi(z), t) = h(z).$$

Note that $\Lambda(\rho, t)$ is a monotone increasing function of t, so the solution exists. Also note that if $D(z)$ is a pseudoconvex point, $D(z) \approx h(z)^{1/2}$, while if $\pi(z)$ is a point of type m, $D(z) \approx h(z)^{1/m}$.

For $\rho \in \partial D$ and $\alpha > 0$ define

$$\mathcal{a}_\alpha(\rho) = \{z \in D \cap W \mid \pi(z) \in B(\rho, \alpha D(z))\}$$

For F holomorphic on D set

$$\mathfrak{M}_\alpha F(\rho) = \sup_{z \in \mathcal{a}_\alpha(\rho)} |F(t)|.$$

We now have an analogue of Theorem A.

Theorem 3: *For* $0 < p \leqslant \infty$ *there are constants* $A_p < \infty$ *so that if* $F \in H^p(D)$,

$$\| \mathfrak{M}_\alpha F \|_{L^p(\partial D)} \leqslant A_p \| F \|_{H^p(D)}.$$

We also have an analogue of Theorem B. There is no general formula for an exact fundamental solution for $\Box_b$, but Rothschild and Stein [5] have constructed an approximate fundamental solution or parametrix for operators of the form $\sum_{j=1}^{p} X_j^2$ and their variants. If $K(x, y)$ is the integral kernel for such a parametrix, we have

Theorem 4:

$$|K(x, y)| \leqslant C \frac{\rho(x, y)^2}{|B(x, \rho(x, y))|}.$$

Complete proofs of theorems 1, 2, and 4 will appear in [4]. Some of this material, as well as Theorem 3, were previously announced in [3].

REFERENCES

[1]. Folland, G. B., and Stein, E. M., *"Estimates for the $\bar{\partial}_b$ Complex and Analysis on the Heisenberg Group"*, **Comm. Pure Appl. Math.** 27 (1974), 429-522.

[2]. Koranyi, A., *"Harmonic Functions on Hermitian Hyperbolic Space"*, **Trans. Am. Math. Soc.** 135 (1969), 507-516.

[3]. Nagel, A., Stein, E. M., and Wainger, S. *"Boundary Behavior of Functions Holomorphic in Domains of Finite Type"*, **Proc. Math. Acad. Sci. U.S.A.** 78 (1981), 6596-6599.

[4]. Nagel, A., Stein, E. M., and Wainger, S. *"Balls and Metrics Defined by Vector Fields I: Basic Properties"*, To appear.

[5]. Rothschild, L. P., and Stein, E. M., *"Hypoelliptic Differential Operators and Nilpotent Groups"*, **Acta Math.** 137 (1976), 247-320.

[6]. Stein, E. M., *Boundary Behavior of Holomorphic Functions of Several Complex Variables*, **Mathematical Notes** #11, Princeton Univ. Press, 1972.

Mathematics Department
University of Wisconsin-Madison
Madison, WI 53706
U.S.A.

R M PORTER

Properties related to directional convexity

In this paper we will describe some geometric conditions, analogous to convexity in a fixed direction, which may be satisfied by a simply connected domain in the plane. The first part contains definitions of several classes of analytic mappings from a disk to such domains (allowing more generally a many-sheeted image) with some coefficient bounds and elementary distortion theory. In the second part we show how each of the geometric conditions may be refined, thus partitioning each class into a two-parameter family of subclasses.

Part I

1.1. We begin by assuming that the domain $D \subseteq \mathbb{C}$ is bounded by a simple analytic curve. Then, for any boundary point $w \in \partial D$, the real quantities

(1) $$\Lambda = \operatorname{Re} w, \quad |w|, \quad \arg w, \quad \arg dw$$

are "geometrically visible." Given a conformal mapping $f: \Delta = \{|z| < 1\} \longrightarrow D$, we will regard each of the quantities Λ as a function of $z = e^{i\theta} \in \partial\Delta$; this is allowable, since by standard results in the theory of prime ends, f extends analytically to the closure of Δ. Let

$$M = \{\text{monotone increasing functions on } [-\pi, \pi]\}$$

and consider the condition $\Lambda \in M$. This can be expressed equivalently by requiring the corresponding θ-derivative

(2) $$\frac{\partial \Lambda}{\partial \theta} = \operatorname{Re} izf', \quad |f| \operatorname{Re} izf', \quad \operatorname{Re} zf'/f, \quad \operatorname{Re}(1+zf''/f')$$

to be non-negative on $\partial\Delta$. By the maximum principle, these harmonic functions will be non-negative inside Δ as well. As is well known, the latter two expressions define the classes of starlike (with respect to 0) and convex functions respectively. The first two expressions describe no non-constant functions, since it is impossible for the bounded function $\operatorname{Re} w$ or $|w|$ to be an increasing and a periodic function of θ at the same time. The observation leads us to replace M with the following class,

$$M' = \{\text{functions non-decreasing on } [-\pi, 0], \text{ non-increasing on } [0, \pi]\}.$$

In Figure 1 are shown some domains for which $\Lambda \in M'$. The first domain is called "vertically convex," since for every vertical line L, the intersection $L \cap D$ is connected. We shall call the remaining

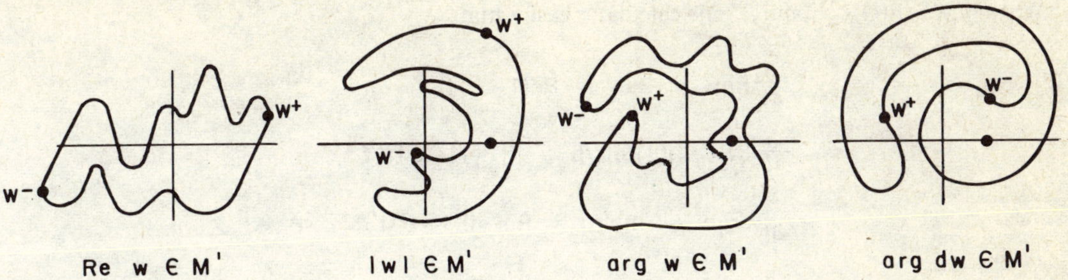

FIGURE 1

three domains "concentrically convex," "radially convex," and "spiral" respectively. Note the positions indicated for the "extreme points" $w^- = f(-1)$, $w^+ = f(1)$.

Vertically convex domains have been studied in [1], [4], [6], [10]. Some analytic functions related to M' (in a slightly different context) appear in [3]. We will now give analytic expressions (5) analogous to (2) for conformal mappings for which $\Lambda \in M'$. For the vertically convex case, the result, as well as our derivation of it, may be found in the paper of Hengartner and Schober [1].

Consider the horizontal band

(3) $$D_0 = \{\zeta = \xi + i\eta\colon |\eta| < \pi/4\}.$$

The function $z = \tanh\zeta$ defines a conformal mapping from D_0 onto Δ. Note that $\partial\theta/\partial\xi > 0$ holds on the "lower boundary" $\eta = -\pi/4$ of D_0, while $\partial\theta/\partial\xi < 0$ holds on the "upper boundary"

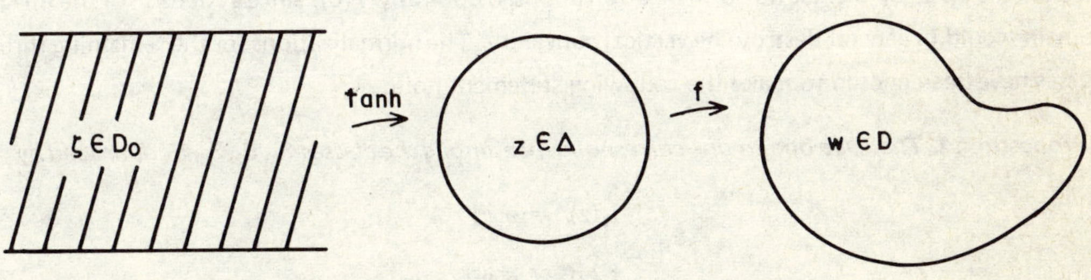

FIGURE 2

$\eta = \pi/4$. Since $\partial\Lambda/\partial\xi = (\partial\Lambda/\partial\theta)(\partial\theta/\partial\xi)$, the condition $\Lambda \in M'$ is equivalent to

(4) $$\frac{\partial \Lambda}{\partial \xi} \geqslant 0.$$

Writing $w = h(\zeta) = f(\tanh \zeta)$ one calculates easily that

(5)
$$\frac{\partial}{\partial \xi} \operatorname{Re} w = \operatorname{Re} h' = \operatorname{Re}(1-z^2) f'$$

$$\frac{\partial}{\partial \xi} |w| = |h| \operatorname{Re} h'/h = |f| \operatorname{Re}(1-z^2) f'/f$$

$$\frac{\partial}{\partial \xi} \arg w = \operatorname{Im} h'/h = \operatorname{Re}(-i)(1-z^2) f'/f$$

$$\frac{\partial}{\partial \xi} \arg dw = \operatorname{Im} h''/h = \operatorname{Re}(-i)((1-z^2) f')'/f'$$

(Regarding the last formula, note that $dw = f'(z)\,dz = \pm h(\zeta)\,d\zeta$ on ∂D_0, where the "+" sign is taken on the lower boundary. Thus, $\arg dw = \arg h'(\zeta)\,d\zeta + C$, and the constant $C = 0$ or π is annihilated by $\partial/\partial\xi$. Further, $\arg d\zeta = 0$ on ∂D_0, so $(\partial/\partial\xi)\arg dw = (\partial/\partial\xi)\arg h' = \operatorname{Im} h''/h'$).

With the preceding considerations as motivation, we introduce the following classes of analytic functions defined in Δ:

(6)
$$S_1 = \{f\colon \operatorname{Re}(1-z^2) f'(z) \geqslant 0,\ f(0) = 0,\ |f'(0)| = 1\}$$

$$S_2 = \{f\colon \operatorname{Re}(1-z^2) f'(z)/f(z) \geqslant 0,\ f(0) = 1,\ |f'(0)| = 1\}$$

$$S_3 = \{f\colon \operatorname{Re}(-i)(1-z^2) f'(z)/f(z) \geqslant 0,\ f(0) = 1,\ |f'(0)| = 1\}$$

$$S_4 = \{f\colon \operatorname{Re}(-i)((1-z^2) f''(z)/f'(z) - 2z) \geqslant 0,\ f(0) = 0,\ f'(0) = 1,\ |f''(0)| = 1\}.$$

There is no longer any requirement that f be analytic on $\partial\Delta$, nor that f be univalent. In these definitions it is implicit that f or f' is not allowed to vanish where it appears in a denominator. Note that for the class S_1 there is no normalization imposed upon any $f'(0)$, since a rotation of the image domain would in general destroy the vertical convexity. The normalizations for the remaining three classes have been chosen to make the following statement hold.

Proposition 1. *There is a one-to-one correspondence among the classes* S_1, S_2, S_3, S_4 *described by*

(7)
$$f_2(z) = e^{f_1(z)}$$

$$f_3(z) = e^{if_1(z)}$$

$$f_4(z) = \int_0^z \frac{e^{if_1(t)}}{1-t^2}\,dt$$

where $f_j \in S_j$, $1 \leqslant j \leqslant 4$.

Note in particular that f_2, f_3, and $(1-z^2)f_4'$ do not vanish in Δ. Thus, they have well-defined holomorphic logarithms. In what follows we will always assume the branch of the logarithm is chosen to vanish at $z=0$. Thus we may write

(8) $$f_1 = \log f_2 = (-i)\log f_3 = (-i)\log(1-z^2)f_4'.$$

1.2. Another class related to S_1 is obtained by noting that if D is allowed to be unbounded, then the first quantity $\Lambda = \mathrm{Re}\, w$ listed in (1) can in fact describe an increasing yet periodic function of θ. When this occurs, D is a vertically convex domain "lying above its boundary," as shown in Figure 3. The corresponding analytic condition $\mathrm{Re}\, izf' \geq 0$ given by (2) may not hold, however, since $\partial\Lambda/\partial\theta$

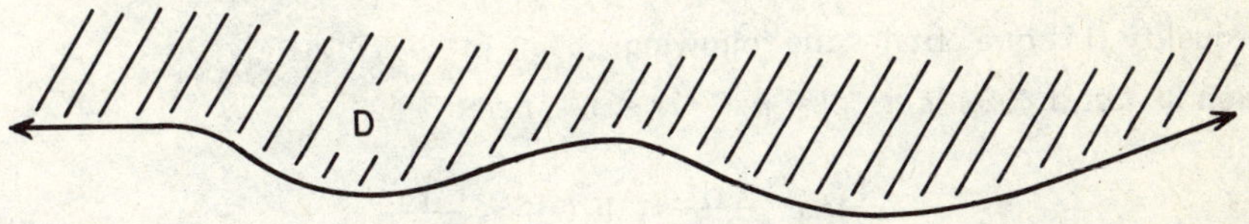

FIGURE 3

blows up at the value of θ corresponding to $w=\infty$. In fact, if $\mathrm{Re}\, izf' \geq 0$ for all $z \in \Delta$, then $\mathrm{Re}\, izf'$ is identically zero, since it vanishes for $z=0$. One may eliminate the troublesome factor z by multiplying by $-(1-z)^2/z$, which is non-negative on $\partial\Delta$. The resulting condition

(9) $$\mathrm{Re}(-i)(1-z)^2 f'(z) \geq 0$$

is that given in [1] characterizing a class of maps to vertically convex domains lying above the boundary. If we replace i by $-i$ in (9), the domain will lie below the boundary. There are also a few special cases such as $f(z) = (1-z)^{-2}$; where all the finite boundary lies to the right or to the left of the domain.

1.3. There are many standard and useful consequences of the inequality $\mathrm{Re}\, p(z) \geq 0$ for $|z| < 1$ (see [7], [12]). For example, applying Schwarz's lemma to $(p(z)-p(0))/(p(z)+\overline{p(0)})$, one obtains

(10) $$|p'(0)| \leq 2\,\mathrm{Re}\, p(0)$$

and

(11) $$\frac{1-|z|}{1+|z|}|p(0)| \leq |p(z)| \leq \frac{1+|z|}{1-|z|}|p(0)|$$

From these inequalities one may deduce facts about the classes defined by (6). For example, (10) translates directly into the following result.

Proposition 2. *Let* $f_j \in S_j$. *Then*

$$|f_1''(0)| \leqslant 2\,\mathrm{Re}\,f_1'(0)$$

$$|f_2''(0) - f_2'(0)^2| \leqslant 2\,\mathrm{Re}\,f_2'(0)$$

$$|f_3''(0) - f_3'(0)^2| \leqslant 2\,\mathrm{Im}\,f_3'(0)$$

$$|f_4'''(0) - f_4''(0)^2 - 2| \leqslant 2\,\mathrm{Im}\,f_4''(0)$$

All of the results of this section relating to the class S_1 may be found in [1]. Note that Proposition 2 is sharp for the function $f_1(z) = z/(1-z)$ and the functions which correspond to it under Proposition 1.

From inequality (11) one obtains the following.

Proposition 3. *Let* $f_j \in S_j$. *Let* $|z| = r$, $0 \leqslant r < 1$. *Then*

$$\frac{1}{2}\log\frac{(1+r)^2}{1+r^2} \leqslant |f_1(z)| \leqslant \frac{r}{1-r}$$

$$e^{-r/(1-r)} \leqslant |f_2(z)| \leqslant e^{r/(1-r)}$$

$$e^{-r/(1-r)} \leqslant |f_3(z)| \leqslant e^{r/(1-r)}$$

$$|f_4(z)| \leqslant e^{r/(1-r)} - 1.$$

Proof. The statement for f_1 is proved in detail in [1]. By (8) it follows that

(12) $$|\log f_2(z)|,\ |\log f_3(z)| \leqslant \frac{r}{1-r}$$

so, in particular

$$|\log|f_2(z)||,\ |\log|f_3(z)|| \leqslant \frac{r}{1-r}$$

which gives the second and third statements. Finally, referring to (8), again we have now

$$\log|(1-z^2)f'(z)| \leqslant \frac{r}{1-r}$$

and deduce that

$$|f_4(z)| = \left|\int_0^z f_4'(t)\,dt\right| \leqslant \int_0^r e^{s/(1-s)}(1-s)^{-1}ds$$

$$= e^{r/(1-r)} - 1,$$

where the integral is over the radial segment from 0 to z.

Missing from Proposition 3 is a lower bound on f_4. One may be derived as follows. By Proposition 1 we take $f_3 \in S_3$ so that

$$f_4(z) = \int_0^z \frac{f_3(t)}{1-t^2}\,dt.$$

Then, by Proposition 3, we have

$$\left|\frac{f_3(t)}{1-t^2}\right| \geqslant a(|t|) \tag{13}$$

where we define

$$a(r) = \frac{1}{1+r^2} e^{-r/(1+r)}. \tag{14}$$

Also, by (12) we have $|\arg f_3(t)| \leqslant |t|(1-|t|)^{-1}$, while for any $t \in \Delta$, $|\arg(1-t^2)| \leqslant \sin^{-1}|t|^2$. Therefore,

$$\left|\arg\frac{f_3(t)}{1-t^2}\right| \leqslant b(|t|), \tag{15}$$

where we define

$$b(r) = \frac{r}{1-r} + \sin^{-1} r^2. \tag{16}$$

Note that $b(r)$ is monotone increasing on $0 \leqslant r < 1$. Define r_0 in this interval by

$$b(r_0) = \pi/2.$$

Then by (13), (15)

$$\operatorname{Re}\frac{f_3(t)}{1-t^2} \geqslant \left|\frac{f_3(t)}{1-t^2}\right| \cos b(|t|) \geqslant a(|t|)\cos b(t),$$

as long as $|t| < r_0$. Since

$$\operatorname{Re}\frac{|z|}{z} f_4(z) = \int_0^z \operatorname{Re}\frac{f_3(t)}{1-t^2}\frac{|z|\,dt}{z}$$

and since $|z|\,dt/z$ is real along the radius from 0 to z, we have

Proposition 4. *Let* $f_4 \in S_4$. *Then*

$$|f_4(z)| \geqslant c(|z|), \qquad |z| \leqslant r_0, \tag{17}$$

where

(18) $$c(r) = \int_0^r a(s)\cos b(s)\,ds.$$

Numerically one finds that $r_0 = .55668$ approximately. In Table 1 we give values of $c(r)$, and for comparison, values of $r/(1+r)^2$ which the Bieberbach distortion theorem specifies as a lower bound for arbitrary normalized univalent functions (recall, however, that f_4 need not be univalent).

r	c(r)	r/(1+r)²
0	0	0
.1	.0943	.0826
.2	.1745	.1389
.3	.2366	.1775
.4	.2766	.2041
.5	.2969	.2222
r_0	.2970	.2297

TABLE 1.

From the lower bounds one obtains the following analogues of the Koebe-Bieberbach "1/4 Theorem".

Proposition 5. *Let* $f_j \in S_j$. *Then* $f_1(\Delta)$ *contains the disk* $|w| < (\log 2)/2$; $f_2(\Delta)$ *and* $f_3(\Delta)$ *contain* $|w - 3\sqrt{2}/4| < \sqrt{2}/4$, *and* $f_4(\Delta)$ *contains* $|w| < c(r_0)$.

Proof. For f_1 we let r tend to 1 in the lower bound of Proposition 3 ([1]). Now, by (8), $f_2(\Delta)$, $f_3(\Delta)$ contain the image of $f_1(\Delta)$ under the exponential mapping. Note that

$$|\exp(\rho e^{i\alpha}) - \cosh\rho|^2 = 2e^{\rho\cos\alpha}\varphi(\alpha) + \sinh^2\rho$$

where $\varphi(\alpha) = \cosh(\rho\cos\alpha) - (\cosh\rho)\cos(\rho\sin\alpha) \geqslant 0$. Therefore, $|\exp(\rho e^{i\alpha}) - \cosh\rho| \geqslant \sinh\rho$, and with $\rho = (\log 2)/2$, one deduces the claims made for f_2, f_3. Finally, the claim for f_4 follows from Proposition 4.

1.4. It is also standard that $\operatorname{Re} p(z) \geqslant 0$ in Δ implies that p admits a Herglotz integral representation

(19) $$p(z) = \int_{-\pi}^{\pi} \frac{p(0)e^{i\theta} + \overline{p(0)}z}{e^{i\theta} - z}\,d\mu(\theta),$$

where μ is a (unique) Borel probability measure on $[-\pi, \pi]$. If we apply this to $f_1 \in S_1$, we obtain

$$(1-z^2)f_1'(z) = \int_\theta \frac{f'(0)e^{i\theta} + \overline{f'(0)}z}{e^{i\theta} - z}\,d\mu(\theta),$$

so that

(20) $$f_1(z) = \int_\theta K_1(z,\theta)\,d\mu(\theta)$$

where

(21) $$K_1(z,\theta) = \int_0^z \frac{f_1'(0)\,e^{i\theta} + \overline{f_1'(0)}\,t}{(1-t^2)(e^{i\theta}-t)}\,dt$$

Formulas for functions in S_2, S_3 in terms of μ are obtained by exponentiating (20). For $f_4 \in S_4$ one finds similarly that

(22) $$\log f_4'(z) = \int_\theta K_4(z,\theta)\,d\mu(\theta),$$

where

(23) $$K_4(z,\theta) = \int_0^z \frac{f''(0)\,e^{i\theta} + (2e^{i\theta} + \overline{f''(0)})\,t - 4t^2}{(1-t^2)(e^{i\theta}-t)}\,dt.$$

1.5. It is observed in [1], [4] that the functions in S_1 are univalent in Δ, since they are close-to-convex; that is, they satisfy $\mathrm{Re}\, f_1'/\varphi' \geqslant 0$, where $\varphi(z) = (1/2)\log(1+z)/(1-z)$ is convex. The functions in S_2, S_3, S_4 are not all univalent.

Proposition 2. *Let* $f \in S_2$ *or* S_3 *and suppose that*

(24) $$\mathrm{Re}\left(\frac{zf'}{f} + \frac{1+z^2}{1-z^2}\right) \geqslant 0$$

for all $z \in \Delta$*. Then* f *is univalent.*

Proof. Define

$$\varphi(z) = \int_0^z \frac{f(t)}{1-t^2}\,dt.$$

Then, by (24), we have $\mathrm{Re}(1+z\varphi''/\varphi') = \mathrm{Re}(zf'/f + (1+z^2)/(1-z^2)) \geqslant 0$; so φ is convex. Since

$$\mathrm{Re}(f'/c\varphi') \geqslant 0,$$

where $c = 1$ or i, we have that f is close-to-convex, hence univalent.

Not every univalent function in S_2, S_3, or S_4 is close-to-convex. This is illustrated by the domain of Figure 4. According to a characterization given by Lewandowski (see [7]), a mapping to D is close-to-convex if and only if the complement of D is a union of rays emanating from ∂D, disjoint from one another, except for these endpoints. This condition clearly fails for the domain shown.

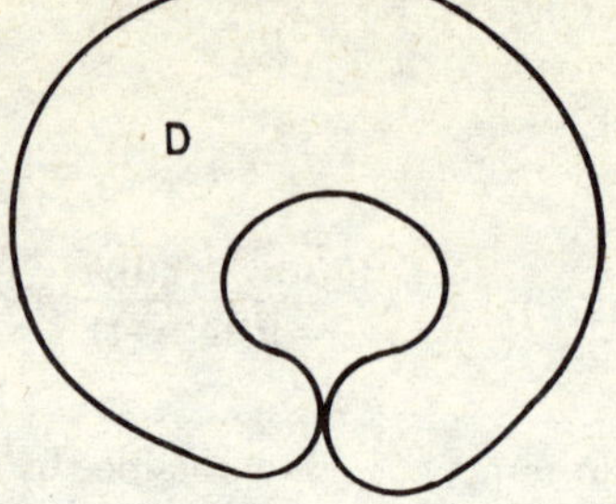

FIGURE 4

It would be of interest to find conditions which make a function in S_4 univalent, and weaker conditions than (24) for S_2, S_3. Also of interest are subclasses obtained by combining the properties we have described, for example, $\{f \in S_1\colon \log f$ is univalent$\}$. Another example is $\{f \in S_3\colon f$ is univalent and for each ray L emanating from 0, $L \cap D$ is connected$\}$. These are the functions which really deserve to be called radially convex; they can be described by the condition that the total variation of $\arg w$ on $\partial\Delta$ does not exceed 2π. Still another example worthy of note is $S_1 \cap S_4$. For this class the upper and lower boundaries of $f(\Delta)$ are graphs of convex functions, so $f(\Delta)$ is a "lune" (Figure 5). It is evident visually that such domains have total boundary rotation (variation of $\arg dw$

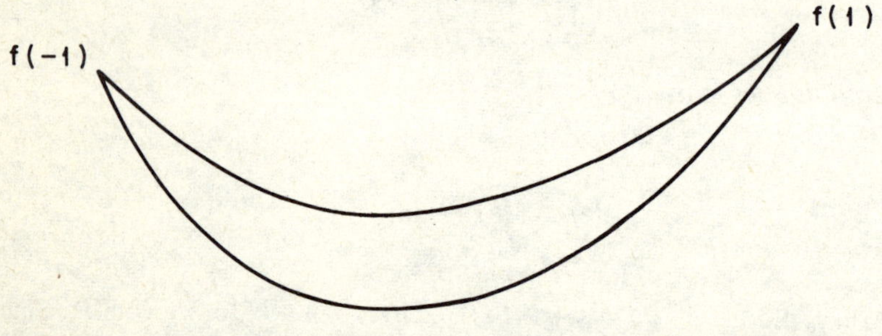

FIGURE 5

on $\partial\Delta$, see [12]) between 2π and 4π. One would expect that it is possible to strengthen the theorems of Section 3 for these kinds of special subclasses; we will not pursue the matter further here.

It happens that none of the inequalities of Section 1.3 is sharp for the subclasses of S_2, S_3, S_4 consisting of univalent functions. For example, suppose that equality were to hold for $f = f_2$ in

Proposition 2. This implies that equality also holds in (10), where $p(z) = (1-z^2)\,f'(z)/f(z)$. According to Schwarz's lemma, we may write $(p(z)-p(0))/(p(z)+\overline{p(0)}) = e^{i\alpha}z$ for some constant α, so

$$(1-z^2)\frac{f'(z)}{f(z)} = \frac{f'(0)+\overline{f'(0)}e^{i\alpha}z}{1-e^{i\alpha}z}.$$

It is not hard to see (dividing by $1-z^2$ and integrating) that $\log f$ is a univalent mapping (of Schwarz-Christoffel type) of Δ onto a vertical band minus a vertical slit (possibly part of the boundary may degenerate according to the values of α and $f'(0)$, leaving just a band or else a slit half plane). As a consequence, f is not univalent.

Now suppose there were univalent functions $F_n \in S_2$ such that

(25) $$\frac{|F_n''(0)-F_n'(0)|}{\operatorname{Re} F_n'(0)} \longrightarrow 2$$

as $n \longrightarrow \infty$. By compactness of the family of normalized univalent functions in Δ, one could then take a subsequence converging to a univalent function f which is easily seen to be in S_2. For $f_2 = f$ equality holds in (10), and we have just seen that this is impossible. This shows that (25) cannot occur; in other words, (10) is not sharp for the univalent subclass of S_2. The other cases are treated similarly.

Part II

2.1. We will now look more closely at the question of the normalization of the mappings. One condition which suggests itself naturally is to prescribe $\arg f'(0)$ in the definition of S_1 and then make the corresponding changes in S_2, S_3, S_4. However, we will look at the problem instead from a different angle to obtain parameters which reflect more precisely the global geometric properties of the image domain.

To begin with, we will assume again for motivation that D has one of the geometric properties we have been discussing and that its boundary is an analytic closed curve. Then D has two extreme points w^-, w^+ as shown in Figure 1 (If D satisfies more than one of the geometric conditions, these extreme points depend on which condition is being considered). Fix $w_0 \in D$ and let $g: \Delta \longrightarrow D$ be the unique conformal mapping satisfying

(26) $$g(0) = w_0, \qquad g'(0) = 0.$$

Since g extends analytically to the closed disk, we may define z^-, z^+ by

(27) $$g(z^-) = w^-, \qquad g(z^+) = w^+.$$

Now let $T: \Delta \longrightarrow \Delta$ be a linear fractional transformation such that

$$T(z^-) = -1, \qquad T(z^+) = 1$$

and define

$$f = g \circ T^{-1},$$

as in Figure 6.

Then, by the analysis given earlier in regard to Figure 2, f satisfies the corresponding property $\Lambda \in M'$, where Λ is one of the expressions in (1). (Note however that f need not satisfy the normalization given in (6).) From this observation it follows readily that g satisfies one of the conditions

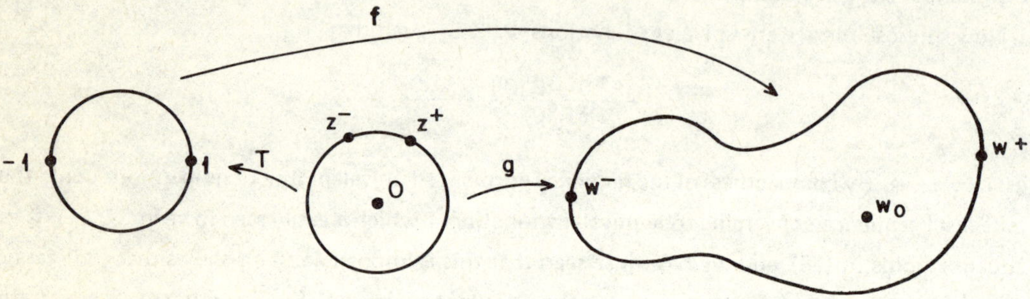

FIGURE 6

(28)
$$\mathrm{Re}\,Pg' \geqslant 0,$$
$$\mathrm{Re}\,Pg'/g \geqslant 0,$$
$$\mathrm{Re}(-i)\,Pg'/g \geqslant 0,$$
$$\mathrm{Re}(-i)(Pg')'/g \geqslant 0,$$

where

(29)
$$P(z) = (-z^- z^+)^{1/2}(z-z^-)(z-z^+)$$

for an appropriately chosen square root of $-z^-z^+$. It is also seen that P does not depend, apart from a positive real multiple, on the choice of T.

Each of the conditions (28) and each pair of values z^-, z^+ determines a particular class of analytic functions in Δ, say, with the normalization (26), or, more conveniently, with the strengthened condition $g'(0) = 1$. For each class there are results analogous to those derived earlier, obtained by exactly the same reasoning. (We have referred to these new classes as "subclasses" of those considered before. To make this precise, we note that the earlier results apply, with inessential changes, to af_1+b: $f_1 \in S_1$, $a \neq 0$ and the classes which correspond under (8)). For the vertically convex case, the resulting theory is particularly interesting; details will appear in [9].

2.2. We close with two remarks. (i) The classes defined by (28) are most similar to those of (6) when $z^- = -z^+$. In fact, for $z^- = -1$, $z^+ = 1$ the defining inequalities are identical. On the other hand, when the parameters merge to a common value on $\partial\Delta$, the image domain becomes of a very special kind. For the case of vertically convex domains, when $z^- = z^+ = 1$, the first inequality of (28) becomes (9). Thus, a vertically convex domain lying above its boundary may be considered as a limiting case of domains having upper and lower boundary, where the upper boundaries have degenerated to a single point at infinity.

(ii) If we allow w_0 to vary in (26), then z^-, z^+ become functions in D. They describe in rough terms how w_0 is situated with respect to the boundary of D. In particular, it can be shown [9] that $\arg z^-$, $\arg z^+$ are harmonic functions in D which can de described conveniently in terms of their brown behavior on the boundary, at least when D is bounded.

REFERENCES

[1] W. Hengartner and G. Schober, *"On Schlicht Mappings to Domains Convex in One Direction,"* **Comm. Math. Helv.** 45 (1970), 303-314.

[2] W. Hengartner and G. Schober, *"A Remark on Level Curves for Domains Convex in One Direction,"* **Applicable Anal.** 3 (1973), 101-106.

[3] W. Kaplan, *"On Gross's Star Theorem, Schlicht Functions, Logarithmic Potentials and Fourier Series,"* **Ann. Acad. Sci. Fenn** AI 86 (1951).

[4] W. Kaplan, *"Close-to-Convex Schlicht Functions",* **Michigan Math. J.** 1 (1952) 169-185

[5] Ch. Pommerenke, *"On Starlike and Close-to-Convex Functions,"* **Proc. London Math. Soc.** 13 (1963), 290-304.

[6] Ch. Pommerenke, *"On Close-to-Convex Analytic Functions,"* **Trans. Am. Math. Soc.** 114 (1965), 176-186.

[7] Ch. Pommerenke, *"Univalent Functions,"* Vandenhoeck and Ruprecht, Gottingen (1975).

[8] R. M. Porter, *"Functions Convex in Two Directions,"* **Bol. Soc. Mat. Mex.** 25 (1980), 59-61.

[9] R. M. Porter, *"Classes of Vertically Convex Domains,"* to appear.

[10] M. Robertson, *"Analytic Functions Starlike in One Direction,"* **Amer. J. Math.** 58 (1936), 465-472.

[11] St. Ruscheweyh and P. Schwittek, *"On Domains Convex in Several Directions,"* preprint.

[12] G. Schober, **Univalent Functions — Selected Topics** (**Lecture Notes in Mathematics** 478), Springer Verlag, Berlin (1975).

Centro de Investigación y Estudios Avanzados del IPN
México, D. F.
México

J RAMANATHAN

Harmonic maps from surfaces to the complex Grassmann manifolds

Introduction

The purpose of this note is to report on progress made in understanding the structure of harmonic maps from S^2 to the Grassmann manifolds and to point out some open questions. In the first section a brief exposition of the structure theorem for harmonic maps of S^2 to $\mathbb{CP}^n$ is given. This was first discovered by the physicists Glaser and Stora [6]. Eells and Wood [4] and Burns [2] have given a rigorous proof of this result. Wolfson [9] has also given a proof from the point of view of moving frames. In the second section we give the outline of the analogous theorem for $G_2(\mathbb{C}^4)$ as target. This was done in [8].

The definition of energy and harmonic maps as well as their basic properties have been omitted here. We refer the reader to [3]. The only point we repeat here is that the energy of a map from a closed surface depends only on its conformal structure. Hence it makes sense to talk of a map from a Riemann surface being harmonic.

Section 1. Harmonic Maps from S^2 to $\mathbb{CP}^n$

Let $h: M \longrightarrow \mathbb{CP}^n$ be a nondegenerate holomorphic map from a given Riemann surface M. Fix a holomorphic chart and a lift $\tilde{h}: U \longrightarrow S^{2m+1}$. For any $p \in U$ define

$$O_k(p) = \operatorname{span}\left\{\frac{d^j \tilde{h}}{dz^j} : 0 \leqslant j \leqslant k\right\}$$

for $k \geqslant 0$ and $O_{-1} = \{0\}$. O_k is independent of the choice of lift and co-ordinate chart. Moreover $\dim O_k = k+1$ everywhere except on a finite set of points. It is well known that O_k extends to a well defined holomorphic map $O_k: M \longrightarrow G_{k+1}(\mathbb{C}^{n+1})$. Set

$$h_k(p) = O_k(p) \cap O_{k-1}^{\perp}(p).$$

$h_k: M \longrightarrow \mathbb{CP}^n$, $k = 0, \ldots, n$ are harmonic with $h_0 = h$ and h_n antiholomorphic.

The crucial point is that for $M = S^2$ the above procedure gives *all* the harmonic maps from S^2 to $\mathbb{CP}^n$.

Theorem 1. [4], [2], [6], [9] *Let* $\varphi: S^2 \longrightarrow \mathbb{CP}^n$ *be a harmonic map that lies in no proper projective subspace of* $\mathbb{CP}^n$. *Then there is a unique nondegenerate holomorphic map* $h: S^2 \longrightarrow \mathbb{CP}^n$ *and an integer* $0 \leqslant k \leqslant n$ *such that* $h_k = \varphi$.

Section 2. Harmonic Maps from S^2 to $G_2(\mathbb{C}^4)$

It is interesting to ask whether there is an analogous theorem for harmonic maps from S^2 to general Grassmann manifolds. This has been done in the special case of $G_2(\mathbb{C}^4)$ in [8]. For full details of what follows we refer the reader to that paper.

Before discussing the result, we introduce some notation. Let

$$E_{k,n} = \{(P, v) \in G_k(\mathbb{C}^n)\times\mathbb{C}^n \colon v \in P\}$$

be the standard vector bundle over $G_k(\mathbb{C}^n)$. Suppose $f\colon M \longrightarrow G_k(\mathbb{C}^n)$ is a C^∞-map from a Riemann surface M. $f^*E_{k,n}$ is a subbundle of $M\times\mathbb{C}^n$ in the natural way. Via this inclusion, $f^*E_{k,n}$ becomes a Hermitian vector bundle with connection. By the theorem of Koszul and Malgrange [7], $f^*E_{k,n}$ can be given a holomorphic structure induced by the (0, 1)-part of the connection. See also [4]. The tangent bundle of $G_k(\mathbb{C}^n)$ is isomorphic to the Hermitian vector bundle $\mathrm{Hom}(E_{k,n}, E^\perp_{k,n})$. The pull-back of the tangent bundle of f can be written as $\mathrm{Hom}(E, E^\perp)$ where $E = f^*E_{k,n}$ and $E^\perp = f^*E^\perp_{k,n}$. The vector bundle $\mathrm{Hom}(E, E^\perp)$ inherits a holomorphic structure from that of E and $E^\perp$. The differential of the map $f\colon M \longrightarrow G_k(\mathbb{C}^n)$ can be interpreted as a section of the bundle $T^*_{\mathbb{C}}M\otimes\mathrm{Hom}(E, E^\perp)$. The splitting $T^*_{\mathbb{C}}M = T^{1,0}M\oplus T^{0,1}M$ of the cotangent bundle gives a decomposition $df = df^+ + df^-$ of the differential of f. The following proposition is standard in the literature. We refer the reader to [4], [5], or [8].

Proposition 2. *The following are equivalent.*

a.) f *is harmonic.*

b.) df^+ *is a holomorphic section in the bundle*

$$T^{1,0}M\otimes\mathrm{Hom}(E, E^\perp)$$

c.) df^- *is antiholomorphic as a section of*

$$T^{0,1}M\otimes\mathrm{Hom}(E, E^\perp).$$

Given a harmonic map $f\colon M \longrightarrow G_k(\mathbb{C}^n)$, the map $f^\perp\colon M \longrightarrow G_{n-k}(\mathbb{C}^n)$ defined by $f^\perp(p) = f(p)^\perp$ is also harmonic. The above preposition then implies that $(df^\perp)^+$ is a holomorphic section of

$$T^{1,0}M\otimes\mathrm{Hom}(E^\perp, E).$$

The composition $(df^\perp)^+\circ df^+$ denotes the section of

$$T^{1,0}M\otimes T^{1,0}M\otimes\mathrm{Hom}(E, E)$$

given by the expression

$$dz\otimes dz\otimes(df^\perp)^+(\tfrac{d}{dz})\circ(df)^+(\tfrac{d}{dz}).$$

It is easy to check that the above expression is a globally-defined, holomorphic section. Now let P be an invariant polynomial of degree r. Then one can check that $P(df^{\perp}(\frac{d}{dz}) \circ df(\frac{d}{dz}))\, dz^{2r}$ is a globally-defined, holomorphic differential on M of degree 2r. If $M = S^2$ any such differential has to vanish because of the Riemann-Roch Theorem. It follows that for $M = S^2$, $df^{\perp}(\frac{d}{dz}) \circ df(\frac{d}{dz})$ is nilpotent for every holomorphic chart (U, z) of S^2.

It turns out that in the special case of $G_2(\mathbb{C}^4)$ the above discussion enables one to classify the harmonic maps from S^2 to $G_2(\mathbb{C}^4)$. Given a harmonic map $f: S^2 \longrightarrow G_2(\mathbb{C}^4)$, the above discussion implies that at least one of $df^{+}(\frac{d}{dz})$ or $(df^{\perp})^{+}(\frac{d}{dz})$ is of rank less than two everywhere. Since both f and $f^{\perp}$ are harmonic maps from S^2 to $G_2(\mathbb{C}^4)$ we may assume that $df^{+}(\frac{d}{dz})$ is of rank less than two for every holomorphic chart (U, z) of S^2. Now, if rank df^{+} is zero everywhere, then $df^{+} \equiv 0$ and f is antiholomorphic. We are interested in understanding the non±holomorphic, harmonic maps. We assume that $df^{+} \not\equiv 0$. It can be argued that the kernel of $df^{+}(\frac{d}{dz})$ gives a globally-defined holomorphic line bundle $\Phi \subseteq E$. Since $E = f^{*}E_{2,4}$ is a subbundle of $S^2 \times \mathbb{C}^4$, Φ can be interpreted as a map $\Phi: S^2 \longrightarrow \mathbb{CP}^3$ with $\Phi(p) \subseteq f(p)$ for all $p \in S^2$. Let Ψ be the orthogonal complement of Φ in E. Ψ also can be interpreted as a map $\Psi: S^2 \longrightarrow \mathbb{CP}^3$. An analysis of the fact that Φ is a holomorphic subbundle of E implies that $\Psi: S^2 \longrightarrow \mathbb{CP}^3$ is harmonic. Theorem 1 can thus be applied to understand Ψ. It turns out that Φ can also be understood in the framework of Theorem 1. This enables one to prove the following two theorems, giving the structure of non±holomorphic maps from S^2 to $G_2(\mathbb{C}^4)$.

Theorem 3. *Let* $\Phi, \Psi: S^2 \longrightarrow \mathbb{CP}^3$ *be maps such that*

a.) $\Psi = h_k$ *for some holomorphic map* $h: S^2 \longrightarrow \mathbb{CP}^3$ *(notation as in Theorem 1)*

b.) $\Phi^{*}E_{1,4} \subseteq ((h_k \oplus h_{k+1})^{\perp})^{*}E_{2,4}$

c.) $\Phi^{*}E_{1,4}$ *is an anti-holomorphic rank one subbundle of* $((h_k \oplus h_{k+1})^{\perp})^{*}E_{2,4}$.

Then the map $f = \Phi \oplus \Psi$ *defined by*

$$f(p) = \Phi(p) \oplus \Psi(p), \qquad p \in S^2$$

is harmonic.

Theorem 4. *If* $f: S^2 \longrightarrow G_2(\mathbb{C}^4)$ *is harmonic and not ±holomorphic, then either* f *or* $f^{\perp}$ *can be written as* $\Phi \oplus \Psi$ *with* Φ *and* Ψ *as above.*

The details of the proofs of the above two theorems beyond what has already been indicated can be found in [8].

Classification theorems of this type are unknown for higher-rank Grassmann manifolds. Progress has been made by Erdem and Wood, who have proved a structure theorem about the so called *isotropic* harmonic maps from S^2 to arbitrary Grassmannians [5]. However, non-isotropic harmonic maps exist even in the case when the target is $G_2(\mathbb{C}^4)$. It is interesting to note that similar questions can be asked for the quaternionic projective spaces $\mathbb{HP}^n$. In the case $\mathbb{HP}^1 \cong S^4$, the harmonic maps from S^2 have been analyzed by Bryant. Using the Penrose twistor fibration he builds a correspondence between holomorphic maps from S^2 to $\mathbb{CP}^2$ and harmonic maps from S^2 to S^4. Actually, his results are much more general than this. We refer the reader to his paper [1] for further information.

REFERENCES

[1] R. Bryant, *"Conformal and Minimal Immersions of Compact Surfaces into the 4-Sphere"*, **Jour. of Diff. Geom.** 17 (1982), p. 455-473.

[2]. D. Burns, *Harmonic Maps from $\mathbb{CP}^1$ to $\mathbb{CP}^n$*, **Springer Lecture Notes in Mathematics** 949 (1982), p. 48-55.

[3]. J. Eells and J. H. Sampson, *"Harmonic Mappings of Riemannian Manifolds"*, **American Journal of Mathematics** 86 (1964), p. 109-160.

[4]. J. Eells and J. Wood, *"Harmonic Maps from Surfaces to Complex Projective Space"*, **Advances in Mathematics** 49 (1983).

[5]. S. Erdem and J. Wood, *"On the Construction of Harmonic Maps into A Grassmannian"*, **J. London Math. Soc.** (2), 28 (1983), p. 161-174.

[6]. V. Glaser and R. Stora, *"Regular Solutions of the $\mathbb{CP}^n$ Model and Further Generalizations"*. Preprint.

[7]. J. L. Koszul and B. Malgrange, *"Sur certaines structures fibrés complexes"*, **Arch. Math.** 9 (1958), p. 102-109.

[8]. J. Ramanthan, *"Harmonic Maps from S^2 to $G_2(\mathbb{C}^4)$"*, **Jour. of Diff. Geom.** (To Appear).

[9]. J. Wolfson, *Minimal Surfaces in Complex Manifolds*, Thesis, Berkeley, (1982).

Department of Mathematics
Ann Harbor, Mich. 48104
U.S.A.

J A SEADE

Vector fields on smoothings of complex singularities

I want to discuss briefly some questions about vector fields on analytic spaces with singularities. I will start by describing the situation for hypersurface germs, where most of these questions are well understood.

Let $f: (\mathbb{C}^{n+1}, 0) \longrightarrow (\mathbb{C}, 0)$ be a holomorphic map with an isolated critical point at $0 \in \mathbb{C}^{n+1}$, and let V be the (singular) variety

$$V = f^{-1}(0),$$

V is a hypersurface of $\mathbb{C}^{n+1}$ with an isolated singularity at 0. We choose $\epsilon > 0$ sufficiently small, and we let M denote the intersection

$$M = V \cap S_\epsilon,$$

where S_ϵ is a sphere in $\mathbb{C}^{n+1}$ of radius ϵ and centre at 0. M is the *link* of 0 in V; it is a smooth (real) manifold of dimension $2n-1$, see [6].

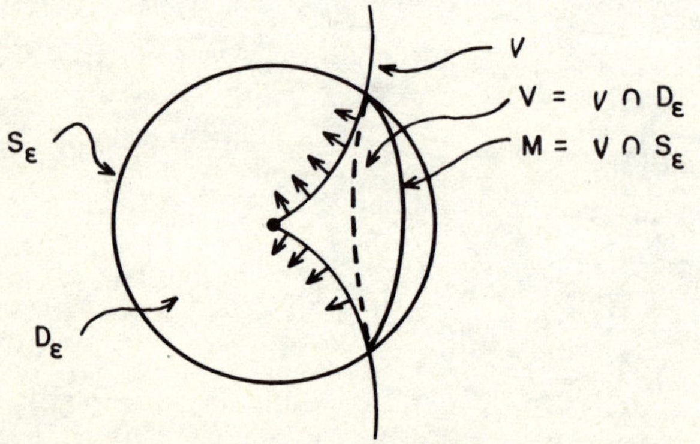

Since by hypothesis 0 is an isolated critical point of f, it follows that if we let $\mathrm{V} = V \cap D_\epsilon$ be the intersection of V with the cloed disc D_ϵ, then the normal bundle of $\mathrm{V}-\{0\}$ is trivialized by the gradient of f. This implies the following facts, as can be seen using standard arguments of topology:

i) $\mathrm{V}-\{0\}$ has (complex) trivial tangent bundle $T(\mathrm{V}-\{0\})$. (Here we mean C^∞-trivialization and not a holomorphic one.)

ii) M is always stably parallelizable, i.e., TM is stably trivial.

Moreover, if we choose $\delta > 0$ small (with respect to ϵ), then for each $t \in \mathbb{C}-\{0\}$ with $|t| < \delta$, the hypersurface $V_t = f^{-1}(f) \subset \mathbb{C}^{n+1}$ is non-singular and intersects S_ϵ transversally.

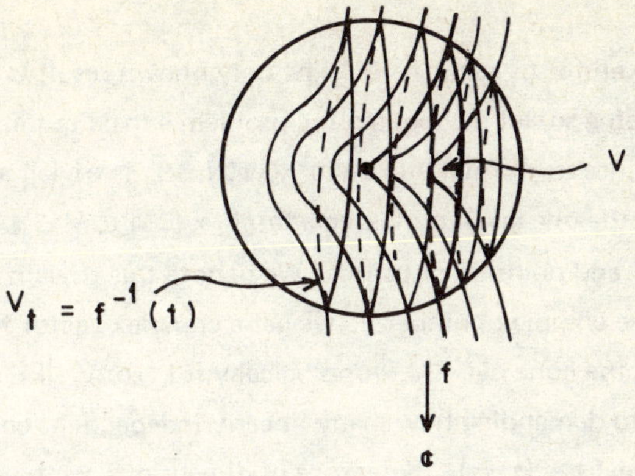

If we let V_t denote the intersection $V_t \cap D_\epsilon$ for $0 < |t| < \delta$, then we have

iii) The complex bundle TV_t is trivial. (V_t is usually called a Milnor fibre of 0.)

We would like to know what happens with i) to iii) when we consider more general singularities not defined by a single equation. (For example, consider the singularities obtained by blowing down to a point, the zero section of the tangent bundle of a Riemann surface of genus $\geqslant 2$.)

Let V be a complex analytic variety of dimension n, with an isolated (normal) singularity at $p \in V$. Think of the germ of V at p as being defined by a set of equations in $\mathbb{C}^m$, and let

$$M = V \cap S_\epsilon$$

denote the intersection of V with a small sphere in $\mathbb{C}^m$ centered at P. M is a smooth (real) submanifold of V of dimension $2n-1$.

Q.1. When is M stably parallelizable? When is it parallelizable?

Remark: If P is a hypersurface germ, M is always stably parallelizable, and if $n \neq 1, 2, 4$, then it is parallelizable *iff* its Milnor number μ is odd [4]. (See [2] or [6] for the definition of μ). For $n = 1, 2, 4$, stably parallelizable implies parallelizable. (This follows from Hirsh's Immersion Theorem, since in these dimensions the sphere S^{2n-1} is parallelizable.)

Q.2. If we let $V = V \cap D_\epsilon$, how many linearly independent complex vector fields are there on $V-\{P\}$?

Remark: If P is a hypersurface germ, then there are n such vector fields on $V-\{P\}$, since $T(V-\{P\})$ is trivial.

A more subtle question is the following,

Q.3. How many linearly independent complex vector fields are there on $V-\{P\}$ that are tangent to the link M?

This question is open even for hypersurfaces. The only known result is the theorem of [5] that we state below. Before doing so, let me explain the problem a little more. M is a codimension 1-real submanifold of V, hence its normal bundle in V, $\nu(N, V)$, is trivial, and we fix a specific trivialization of this bundle: the one given by taking at each $x \in M \subset V \subset \mathbb{C}^m$ a unit vector tangent to V at P, normal to M and pointing outwards. We denote this trivialization of $\nu(N, V)$ by τ. Upon multiplication by the complex number i, we get a complex vector field $(\tau, i\tau)$ on V over M. [We remark that V is the cone over M, topologically [6], so $V-\{P\}$ is $M \times \mathbb{R}$.] The question above is then equivalent to demanding how many linearly independent complex vector fields can we construct on M normal to $(\tau, i\tau)$. By reasons of dimensions, we note that this number is at most $(n-1)$. When is it precisely $(n-1)$? Call this number $\pi(P)$.

Theorem [5]. *Let (V, P) be the germ of a hypersurface singularity in $\mathbb{C}^{n+1}$, $n \geq 1$, and let μ be the Milnor number of* P. *Then $\pi(P)$ is $(n-1)$ if and only if*

$$1+(-1)^n \mu \equiv 0 \qquad \operatorname{mod}(n-1)!$$

Outline of the Proof. We first recall [6] that if we let $F \equiv V_t$ be a Milnor fibre of P, then F is a parallelizable manifold with the homotopy type of a wedge of n-spheres, the number of spheres in this wedge being the number μ. So the Euler characteristic of F is $\chi(F) = 1+(-1)^n\mu$. Now, F has a complex structure on its interior, and its boundary is (diffeomorphic to) the link M. Thus, the theorem above is equivalent to the following statement: there exists a C^∞-trivialization of the complex bundle $TF|M$ given by sections $(\alpha_1, \dots, \alpha_n)$ with the real part of α_1 being normal to M in F *iff* $\chi(F) = m\cdot(n-1)!$ for some $m \in \mathbb{Z}$. This is the statement that we prove. For this we make two observations:

a) Every complex trivialization $\beta = (\beta_1, \dots, \beta_n)$ of $TF|M$ has a degree $d(\beta) \in \mathbb{Z}$, which may be defined as the degree of the map $M = \partial F \longrightarrow S^{2n-1}$ given by

$$X \longrightarrow \beta_i(X) \in TF|M \subset T\mathbb{C}^{n+1}|M$$

for some (anyone) i.

b) The degree of a complex trivialization of $T\mathbb{C}^n|S^{2n-1}$ is a multiple of $(n-1)!$ Indeed, all possible multiples occur. (This observation follows because the map $\pi_{2n-1}(U(n)) \longrightarrow \pi_{2n-1}(S^{2n-1})$ is multiplication by $(n-1)!$)

Now, if $\chi(F) = m\cdot(n-1)!$ for some $m \in \mathbb{Z}$, then we can always construct a trivialization $\beta =$

$(\beta_1, \dots \beta_n)$ of $TF|M$ with degree $m\cdot(n-1)!$, so $d(\beta) = \chi(F)$. Therefore β is homotopic to a trivialization $TF|M$ as stated. The converse is more difficult; it has two main steps:

i) We show that given a trivialization β of $TF|M$ we can always find a torus $T \cong S^n \times D^n$ embedded in the interior of F, such that β extends to $F - \text{Int}\, T$.

From this one concludes

ii) The degree of any such β as in i) is a multiple of $(n-1)!$

Therefore, if we let $\alpha = (\alpha_1, \dots, \alpha_n)$ be a trivialization of $TF|M$ with the real part of α_1 being normal to M, then the degree of α is $\chi(F)$, by Hopf's Theorem for manifolds with boundary. Hence $\chi(F)$ is a multiple of $(n-1)!$

Let me now introduce some common notation (see [1, 2, 3]). Let V be a complex variety of dimension n, with a normal isolated singularity at $P \in V$, and let W be a complex variety of dimension $(n+1)$, containing V as a subvariety. (Everything is germ). A holomorphic map

$$F: W \longrightarrow \Delta \subset \mathbb{C}$$

from W into an open disc Δ in $\mathbb{C}$, is a *smoothing* of P if the following hold:

i) $F^{-1}(0) = V$.

ii) $F^{-1}(t) = V_t$ is non-singular for all $t \in \Delta - \{0\}$.

iii) F is flat at P, i.e., F is not a zero divisor in the local ring $O_{W,P}$.

Under these circumstances, W is either regular at P, or else it has been an isolated, normal singularity there. If it is regular at P, then its germ at P is that of $\mathbb{C}^{n+1}$, and we are in the situation envisaged before, where the germ of V at P was a hypersurface germ. In any case, we may think of W as embedded in some complex space $\mathbb{C}^m$, then consider a sufficiently small sphere S_ϵ in $\mathbb{C}^m$ around P so that $M = V \cap S_\epsilon$ is the link of P, and for each $t \in \Delta - \{0\}$ sufficiently small, the manifold V_t intersects S_ϵ transversally. Following [2], let us call the intersection $V_t = V \cap D_\epsilon$ a Milnor fibre (of the smoothing) of P.

Q.4. How many linearly independent complex vector fields are there on V_t?

Let us denote this number by $X(P)$. Does the number $X(P)$ depend on the smoothing? If we denote by $Y(P)$ the number of linearly independent complex vector fields on $V - \{0\}$, then clearly $X(P) \leqslant Y(P)$. When do we have equality?

The question below might be of interest for homotopy theory.

Q.5. Does there exist a singularity (V, P) as before, such that the link M is stably parallelizable, but M is not the boundary of any compact, parallelizable manifold?

If such a manifold exists, it would be a good candidate for producing element in the stable homotopy of spheres that are outside of the image of the bi-stable J-homomorphism, see [8].

The following proposition of the author answers the above questions in a special case. (The first part of the proposition is implicit in [1]; the last statement was conjectured by A. Durfee in [1; 1.6].)

Proposition [7]. *Let V be 2-dimensional and assume P is also Gorenstein. Then the bundle* $T(V-\{0\})$ *is (complex) trivial. Moreover, if* P *is smoothable and* V_t *is the Milnor fibre of a smoothing of* P, *then* TV_t *is also (complex) trivial.*

We recall that P is *Gorenstein* if there exists a nowhere-vanishing holomorphic 2-form on a punctured neighbourhood of P in *V*. A good account of Gorenstein singularities is given in [3].

Proof. The 2-dimensional vector bundles over $V-\{0\}$ are classified by a map $V-\{0\} \longrightarrow BU(2)$, where BU(2) is the classifying space of the 2-dimensional complex bundles. But $V-\{0\} \cong M \times \mathbb{R}$, so it has the homotopy of a CW-complex of dimension 3 and therefore every map into BU(2) lifts to BU(1). Hence, these vector bundles over $V-\{0\}$ are classified by their first Chern class. Thus, the result will follow if we prove that $C_1(T(V-\{0\}))$ vanishes. Since V is 2-dimensional,

$$C_1(T(V-\{0\})) = -K$$

where K is the Chern class of the canonical bundle $\Lambda^2 T^*(V-\{0\})$. Thus the first part of the proposition follows because $\Lambda^2 T^*(V-\{0\})$ is trivial, since P is Gorenstein.

Using similar arguments, we obtain the rest of the proposition as a consequence of the following lemma.

Lemma. *The canonical bundle* $\Lambda^2 T^*$ *of the interior of* V_t *is holomorphically trivial.*

Proof of Lemma. Here, by interior of V_t we mean $V_t - \partial V_t \cap \mathring{D}_\epsilon$.

Now, let

$$\mathcal{O}_{V,P} \cong \mathcal{O}_{W,P/(F)}$$

be the local ring of *V* at P. Since P is Gorenstein, (*) there are elements $\overline{g}_1, \overline{g}_2 \in \mathcal{O}_{V,P}$ which form a regular sequence (i.e., $\overline{g}_1$ is not a zero-divisor in $\mathcal{O}_{V,P}$ and $\overline{g}_2$ is not a zero divisor in $\mathcal{O}_{V,P}/(\overline{g}_1)$), and such that

$$\mathrm{Hom}_{\mathcal{O}_{V,P}}\{\mathbb{C}, \mathcal{O}_{V,P}/(\overline{g}_1, \overline{g}_2)\} \cong \mathbb{C}$$

Hence, if $g_1, g_2 \in \mathcal{O}_{W,P}$ are extensions of $\overline{g}_1, \overline{g}_2$, (which exists by normality), then (f, g_1, g_2) form

(*) I wish to thank M.V. Nori for explaining me these arguments.

a regular sequence in $O_{W,P}$ and

$$\mathrm{Hom}_{O_{W,P}}\{\mathbb{C}, O_{W,P}/(f, g_1, g_2)\} \cong \mathbb{C}$$

hence W is Gorenstein at P, see [3]. Thus, there exists a nowhere-vanishing holomorphic 3-form ω on a punctured neighbourhood of P in W.

Since F is flat at P, we have an exact sequence of vector bundles over $W-V$,

$$0 \longrightarrow T_F \longrightarrow T(W-V) \longrightarrow F^*(T\Delta-\{0\}) \longrightarrow 0$$

where T_F is the tangent bundle along the fibers of F, i.e. the kernel of dF. Hence

$$\Lambda^3 T^*(W-V) \cong \Lambda^2 T_F^* \otimes F^* T^*(\Delta-\{0\})$$

and $T^*(\Delta-\{0\})$ is naturally trivialized by the form dz on $\mathbb{C}$. Hence the above 3-form ω defines a nowhere-vanishing holomorphic 2-form along each fibre of F, and the lemma follows.

As a consequence of the proposition above, we see that for Gorenstein surface singularities, the number of vector fields as in Q.2 is two. If the singularity is smoothable, then the number of vector fields as in Q.4 is also two, independently of the smoothing. Also, for these singularities the number $\pi(P)$ of Q.3 is 1, i.e., we may always construct three linearly independent real vector fields on M, say ρ_1, ρ_2 and ρ_3, such that at each $x \in M$ we have $\rho_3(x) = i\rho_2(x)$ and $(-i)\rho_1(x)$ is the unit outwards normal field of M in V, see [7].

I thank the referee for his comments.

REFERENCES

[1]. A. Durfee, *"The Signature of Smoothings of Complex Surface Singularities"*, **Math. Annalen** 232, (1978), p. 85-98.

[2]. G. M. Greuel & J. Steenbrink, *"On the Topology of Smoothable Singularities"*, **AMS-Proc. Symp. Pure Maths. No.** 40, Part 1, (1983), p. 535-545.

[3]. R. Hartshorne, *Residues and Duality.* **Lecture Notes in Maths. No.** 20, Springer Verlag, 1966.

[4]. Li Banghe, *"Parallelizable Manifolds and the Stable Homotopy of Spheres".* Preprint.

[5]. Li Banghe, & J. A. Seade, *"Framings on Algebraic Knots".* Preprint.

[6]. J. Milnor, *Singular Points of Complex Hypersurfaces,* **Ann. of Math. Studies** #61. Princeton Univ. Press, 1968.

[7]. J. A. Seade, *"A cobordism Invariant for Surface Singularities"*, **AMS-Proc. Symp. Pure Maths. No.** 40, Part 2, (1983), p. 479-484.

[8]. K. Knapp, *"On the Bi-Stable J-Homomorphism"*, **Lecture Notes in Mathematics** #763, Springer Verlag, 1979. **Algebraic Topology,** Aarhus 1978. Edited by J. L. Dupont and I. H. Madsen.

INSTITUTO DE MATEMATICAS
UNIVERSIDAD NACIONAL AUTONOMA DE MEXICO

M J SPURR

On the genus of an irreducible component of the zero set of a holomorphic one-form

Introduction

The zero set of a holomorphic vector field on a compact complex manifold M has been known for some time to convey structural information concerning M. In addition to Gauss-Bonnet-type results, the Carrell-Lieberman Theorem [1] relates the dimension of the zero set of a holomorphic vector field on a compact Kähler manifold M to cohomological vanishing. More precisely, if M has a holomorphic vector field with zero set of dimension k, then $H^q(M, \Omega_M^p) = 0$ for $|p-q| > k$ (where Ω_M^p is the sheaf of germs of holomorphic p-forms on M).

In the dual case, the zero set of a holomorphic 1-form has been found to convey structural information for a compact complex surface S. Using an idea of A. J. Sommese to utilize the Albanese mapping and the characterization of an exceptional set in S in terms of the negative definiteness of its self intersection matrix, the author has proven [7, 8] the following

Theorem 1: *Let* S *be a compact complex surface having a holomorphic 1-form* ϕ*. If* D *is any divisor on* S *satisfying* $D \cdot D \geqslant 0$ *and* ϕ *pulls back to zero on* D*, then:*
There exists a holomorphic map $f: S \longrightarrow R$ *onto a compact Riemann surface satisfying:*

(1) $\phi = f^*(\phi_R)$ *for some* $\phi_R \in H^0(R, \Omega_R^1)$.

(2) f *has connected fibers.*

(3) Each component of D *is setwise-contained in a fiber of* f.

(4) Each connected component of D *is a rational multiple of the natural divisor associated to the fiber of* f *containing it, and so* $D \cdot D = 0$.

Notationally, $D \cdot D$ is the intersection pairing of D with itself. Also, given a divisor $D = \sum_{j=1}^{J} n_j D_j$, and letting $i_j: D_j \longrightarrow S$ be inclusion mappings, then ϕ is said to pull back to zero on D iff $i_j^* \phi = 0$ on D_j for each j.

If a holomorphic 1-form vanishes or more generally pulls back to zero on a divisor D, then Theorem 1, Part (4) rules out positivity of $D \cdot D$. This yields:

Corollary 1: *If* S *is a compact complex surface and* $\phi \in H^0(S, \Omega_S^1)$ *pulls back to zero on* D*, then* $D \cdot D \leqslant 0$.

This article will use the above results to show that the genus g(C) of an irreducible component C of the zero set of a holomorphic 1-form on a compact surface S is bounded above in terms of Chern numbers of S. As is customary, $g(C)$ is taken to be the genus of the normalization of C if C is singular. The main result is:

Theorem 2: *Let S be a compact complex surface and let C be an irreducible curve contained in the zero set of a holomorphic 1-form ϕ on S. Then the genus of C is bounded by*

$$g(C) \leqslant 1 + \frac{3}{2}\max(0, c_2(S)).$$

There are also sharper bounds on $g(C)$ depending on the category of the surface S. These are given in:

Theorem 3: *Let S be a compact surface free from exceptional curves, and let C be an irreducible curve contained in the zero set of a holomorphic 1-form ϕ on S. Let $h^{1,0} = \dim H^0(S, \Omega^1_S)$, $H_\phi = \{\psi \in H^0(S, \Omega^1_S) \mid \psi \wedge \phi \equiv 0\}$, and $h_\phi = \dim H_\phi$.*
Then $g(C)$, the genus of C, is bounded by one of the following

(1) $g(C) = 0$ *if S is ruled*

(2) $g(C) \leqslant 1$ *if S is elliptic*

(3) $g(C) \leqslant 1 + \frac{1}{2}c_1^2(S)$ *if $h^{1,0} \geqslant 2$ and there exists $\psi \in H^0(S, \Omega^1_S)$ with $\phi \wedge \psi \not\equiv 0$*

(4) $g(C) \leqslant 1 + \dfrac{c_2(S)}{4(h_\phi - 1)} \leqslant 1 + \frac{1}{4}c_2(S)$ *if $h^{1,0} \geqslant 2$, S is not ruled, and $h_\phi \geqslant 2$ (there exists $\psi \in H^0(S, \Omega^1_S)$ with $\phi \wedge \psi \equiv 0$).*

(5) $g(C) \leqslant 1 + \frac{1}{2}c_2(S)$ *if $h^{1,0} = 1$ and S is neither elliptic nor ruled.*

Hence, $g(C) \leqslant 1 + \frac{1}{2}\max(0, c_1^2(S), c_2(S))$.

The proof of the theorems will rely on the fact that $g(C) \leqslant \pi(C)$, where $\pi(C) = 1 + \frac{1}{2}(K \cdot C + C \cdot C)$ is the virtual genus of C and K is the canonical divisor of the surface containing C. Corollary 1 gives $C \cdot C \leqslant 0$ for C contained in the zero set of a holomorphic 1-form. So $g(C) \leqslant 1 + \frac{1}{2}K \cdot C$ and it is only necessary to bound $K \cdot C$.

Section 1: Basic Lemmas

We mention some facts about the behaviour of the canonical divisor K.

Lemma 1: *If S is a compact surface free from exceptional curves, with $p_g(S) = h^{2,0} > 0$, and if K_j is an irreducible component of K, then $K \cdot K_j \geqslant 0$ and hence, $K \cdot K = c_1^2(S) \geqslant 0$.*

Proof: See [2].

Lemma 2: *If S is a non-ruled compact surface free from exceptional curves and if $f: S \longrightarrow R$ is holomorphic onto a Riemann surface with f having connected fibers, then $K \cdot C \geqslant 0$ for C an* *irreducible component of a fiber of* f.

Proof: $\pi(C) \geqslant 0$ gives $K \cdot C \geqslant -C \cdot C - 2$. By a theorem of Zariski [6] $C \cdot C \leqslant 0$. If $C \cdot C \leqslant -2$, then $K \cdot C \geqslant 0$ as desired. If $C \cdot C \leqslant -1$ and $K \cdot C < 0$, then C is rational exceptional contradicting the hypothesis. If $C \cdot C = 0$ and $K \cdot C < 0$ then $K \cdot C = -2$. So $\pi(C) = 0$ and C is a smooth rational fiber of f. If C has multiplicity $m \geqslant 1$ and if F is a regular fiber of f, then $\pi(mC) = \pi(F) \geqslant 0$. So $1 + \frac{1}{2}(mC \cdot K + m^2 C \cdot C) = 1 + \frac{1}{2}(m(-2) + 0) = 1 - m \geqslant 0$ gives that $m = 1$ and that C is a smooth regular fiber. S is then ruled, contrary to assumption. So $K \cdot C \geqslant 0$ as claimed.

Lemma 3: *If* S *is a compact surface free from exceptional curves and if* $f: S \longrightarrow R$ *is holomorphic onto a Riemann surface with* f *having connected fibers, then* $\pi(C) \leqslant \pi(F)$ *for* C *an irreducible curve contained in a fiber of* f *and for* F *a regular fiber of* f.

Proof: If S is ruled $\pi(C) = 0 = \pi(F)$; if S is not ruled, and if $\overline{F} = \sum_{j=1}^{J} n_j F_j$ is the divisor associated to the fiber of f containing C, then $F \cdot F = \overline{F} \cdot \overline{F} = 0$ and $K \cdot F = K \cdot \overline{F}$. By lemma 2, $K \cdot F_j \geqslant 0$ for each $F_j \subseteq \overline{F}$, hence,

$$\begin{aligned}\pi(F) &= 1 + \frac{1}{2}(K \cdot F + F \cdot F) \\ &= 1 + \frac{1}{2}(K \cdot \overline{F}) = 1 + \frac{1}{2}\sum_{j=1}^{J} n_j F_j \cdot K \\ &\geqslant 1 + \frac{1}{2} K \cdot C \geqslant 1 + \frac{1}{2}(K \cdot C + C \cdot C) = \pi(C)\end{aligned}$$

The last inequality follows since $C \cdot C \leqslant 0$ by the theorem of Zariski [6].

Section 2: Bounding the Genus

In this section the proofs of Theorem 3 and Theorem 2 are provided.

Proof of Theorem 3: If S is either elliptic or ruled and has a holomorphic 1-form ϕ which vanishes on C, then ϕ is a pullback of a holomorphic 1-form on the base curve and C is contained in a fiber of the elliptic or ruling map. Hence, $\pi(C) = g(C) = 0$ if S is ruled giving (1), and $\pi(C) \leqslant 1$ if S is elliptic, giving (2).

(3) If $h^{1,0} \geqslant 2$ and there is a $\psi \in H^0(S, \Omega_S^1)$ with $\phi \wedge \psi \not\equiv 0$ then $p_g(S) > 0$ and C is contained in the zero divisor $K = \sum_{j=1}^{J} n_j K_j$ of $\phi \wedge \psi$, a canonical divisor. So $K \cdot K = \sum_{j=1}^{J} n_j K \cdot K_j \geqslant K \cdot C$ as $K \cdot K_j \geqslant 0$. Since $C \cdot C \leqslant 0$ by Corollary 1, $K \cdot K \geqslant K \cdot C + C \cdot C$ and $g(C) \leqslant \pi(C) = 1 + \frac{1}{2}(K \cdot C + C \cdot C) \leqslant 1 + \frac{1}{2} K \cdot K$ giving the bound for this case.

(4) If $h^{1,0} \geqslant 2$ and $h_\phi \geqslant 2$ (that is, there exists $\psi \in H^0(S, \Omega_S^1)$ with $\phi \wedge \psi \equiv 0$ and $\psi \neq k\phi$ for any $k \in \mathbb{C}$), then by a theorem of Castelnuovo-De Franchis [10] there exists $f: S \longrightarrow R$ a holomorphic map onto a Riemann surface of genus 2 or more, with f having connected fibers and with each of ϕ and ψ being a pullback of a holomorphic 1-form on R.

For the moment assume $h_\phi = g(R)$. The map $f: S \longrightarrow R$ gives by a theorem of Campedelli [10] that

$c_2(S) \geq 4(g(F)-1)(g(R)-1)$ where F is a regular fiber of f. Since S is not ruled and $g(R) = h_\phi \geq 2$, we have $c_2(S) \geq 0$ and

$$g(F) \leq 1 + \frac{c_2(S)}{4(g(R)-1)} = 1 + \frac{c_2(S)}{4(h_\phi - 1)}.$$

Now C is contained in a fiber of f since ϕ vanishes on C, giving $\pi(C) \leq \pi(F)$. So

$$g(C) \leq \pi(C) \leq \pi(F) = g(F) = 1 + \frac{c_2(S)}{4(h_\phi - 1)} \leq 1 + \frac{c_2(S)}{4}$$

giving the desired bound for (4).

To show $g(R) = h_\phi$ it will be shown that $f^*(H^0(R, \Omega^1_R)) = H_\phi$. Clearly $f^*(H^0(R, \Omega^1_R)) \subseteq H_\phi$ as $\phi \in f^*(H^0(R, \Omega^1_R))$. Conversely, let $\xi \in H_\phi$. About each $p \in S' = S \setminus$(non-regular fibers of f) there is a coordinate neighborhood with coordinates (z_1, z_2) such that z_1 is a coordinate about $f(p)$ in R. Now $\phi \in f^*H^0(R, \Omega^1_R)$ gives $\phi = a(z_1)dz_1$ locally. Writing $\xi = b(z_1, z_2)dz_1 + e(z_1, z_2)dz_2$ and using the fact that $\phi \wedge \xi \equiv 0$, gives $e(z_1, z_2) \equiv 0$ and $\xi = b(z_1, z_2)dz_1$. For fixed z_1, $b(z_1, z_2)$ induces a holomorphic function on each regular fiber, giving that $b(z_1, z_2) = b(z_1)$ and $\xi = b(z_1)dz_1$. So ξ on S' is a pullback of a holomorphic 1-form $\overline{\xi}$ on $R\setminus\{a_1, \ldots, a_k\}$ where the a_j are the non-regular points of f in R. About a smooth point p in a non-regular fiber, select coordinates (u_1, u_2) centered at p with $z_1 = u_1^n$ a coordinate on R centered at f(p). Now $\overline{\xi} = b(z_1)dz_1$ with $b(z_1) = \sum_{j=-\infty}^{\infty} b_j z^j$ for $b_j \in \mathbb{C}$. So $\xi = b(u_1^n)nu_1^{n-1}du_1$ away from $u_1 = 0$. This, along with the fact that ξ is holomorphic, implies $b(z_1) = \sum_{j=0}^{\infty} b_j z^j$ is holomorphic at $z_1 = 0$. So $\overline{\xi}$ extends to $\xi_R \in H^0(R, \Omega^1_R)$ and $\xi = f^*\xi_R$ giving that $H_\phi = f^*H^0(R, \Omega^1_R)$.

(5) If $h^{1,0} = 1$ and S is neither elliptic nor ruled, then S must be Kähler. The Albanese mapping $\alpha_S: S \longrightarrow T$ maps S onto a 1-dimensional torus. C is contained in a fiber of α_S, α_S has connected fibers, and $\phi = \alpha_S^*(\phi_T)$ for ϕ_T the holomorphic 1-form on T.

Let $D_\phi = \sum_{j=1}^{J} n_j D_j$ be the zero divisor associated to ϕ. The D_j are contained in fibers of α_S; let $D_1 = C$. Let L be the line bundle associated to D_ϕ, and λ be the tautological section of L. For T^*S the holomorphic cotangent bundle of S, $T^*S \otimes L^{-1}$ has a holomorphic section with isolated zeroes, namely $\phi \otimes \lambda^{-1}$. Hence, $c_2(T^*S \otimes L^{-1}) = D_\phi^2 - K \cdot D_\phi + c_2(S) \geq 0$.

So

$$\begin{aligned} c_2(S) &\geq -D_\phi^2 + K \cdot D_\phi \\ &\geq K \cdot D_\phi \quad (\text{as } D_\phi^2 = D_\phi \cdot D_\phi \leq 0 \text{ by Corollary 1}) \\ &= \sum_{j=1}^{J} n_j K \cdot D_j \end{aligned}$$

Lemma 2 gives that $K \cdot D_j \geq 0$ for each j, and so $c_2(S) \geq \Sigma\, n_j K \cdot D_j \geq K \cdot D_1 = K \cdot C$. This gives, since $C \cdot C \leq 0$, that

$$g(C) \leqslant \pi(C) = 1+\frac{1}{2}(K\cdot C+C\cdot C) \leqslant 1+\frac{1}{2}K\cdot C \leqslant 1+\frac{1}{2}c_2(S)$$

which is the desired bound for (5).

Proof of Theorem 2: Blow down S, freeing it of exceptional curves. Let the resulting surface be $\overline{S}$, the resulting curve be $\overline{C}$, and the induced 1-form be $\overline{\phi}$. (If $\overline{C}$ were a point, then $g(C) = 0$). By Theorem 3,

$$g(C) = g(\overline{C}) \leqslant 1+\frac{1}{2}\max(0, c_1^2(\overline{S}), c_2(\overline{S})).$$

Consulting Kodaira's classification of compact complex surfaces [4], and using the fact that $\overline{S}$ has a holomorphic 1-form $\overline{\phi}$ vanishing on $\overline{C}$, $\overline{S}$ must lie in one of the following classes of surfaces free from exceptional curves:

I Algebraic surfaces with $p_g = 0$

IV Elliptic surfaces with b_1 even, $p_g > 0$, $c_1^2 = 0$, $c_1 \neq 0$

V Algebraic surfaces with $p_g > 0$ and $c_1^2 > 0$

VI Elliptic surfaces with b_1 odd, $p_g > 0$, and $c_1^2 = 0$

If $\overline{S}$ lies in Class I, then it must be ruled or elliptic [3, 9]. If $\overline{S}$ lies in Class IV or VI, it is elliptic. If $\overline{S}$ is either ruled or elliptic, $\overline{\phi}$ is a pullback of a holomorphic 1-form on the base curve and $\overline{C}$ is contained in a fiber. Hence, $g(C) = g(\overline{C}) \leqslant 1 \leqslant 1+\frac{3}{2}\max(0, c_2(S))$.

If $\overline{S}$ lies in Class V, then it is algebraic of general type. Miyaoka [5] has shown that for surfaces of general type $c_1^2 \leqslant 3c_2$. Hence,

$$\begin{aligned} g(C) = g(\overline{C}) &\leqslant 1+\frac{1}{2}\max(0, c_1^2(\overline{S}), c_2(\overline{S})) \\ &\leqslant 1+\frac{3}{2}\max(0, c_2(\overline{S})) \\ &\leqslant 1+\frac{3}{2}\max(0, c_2(S)) \end{aligned}$$

as c_2 increases by 1 each time a surface is blown up.

REFERENCES

[1]. Carrell, J.. B. and Lieberman, D. I., *Holomorphic Vector Fields and Compact Kaehler Manifolds,* **Invent. Math.** 21 (1973), 303-309.

[2]. Kodaira, K., *On the Structure of Compact Complex Analytic Surfaces I,* **Amer. J. Math.** 86 (1964), 751-798.

[3]. Kodaira, K., *On the Structure of Compact Complex Analytic Surfaces IV,* **Amer. J. Math.** 90 (1968), 1048-1066.

[4]. Kodaira, K., *On the Structure of Compact Complex Analytic Surfaces II,* **Proc. Nat. Acad. Sci. U.S.A.** 51 (1964),

1100-1104.

[5]. Miyaoka, Y., *On the Chern Numbers of Surfaces of General Type,* **Invent. Math.** 42 (1977), 225-237.

[6]. Safarevich, I, et al., *Algebraic Surfaces,* **Proc. Steklov Inst. of Math., Am. Math. Soc.,** Providence, R.I. (1967).

[7]. Spurr, M.J., *On the Zero Set of A Holomorphic One-Form,* Thesis, Tulane University, August, 1983.

[8]. Spurr, M. J., *On the Zero Set of A Holomorphic 1-Form on A Compact Complex Surface,* To Appear.

[9]. Ueno, K., *Classification Theory of Algebraic Varieties and Compact Complex Spaces,* **Lecture Notes in Mathematics,** 439, Springer Verlag (1975).

[10]. Van de Ven, A., *On the Chern Numbers of Surfaces of General Type,* **Inv. Math.** 36 (1976), 285-293.

Department of Mathematics
Tulane University
New Orleans, LA 70118
U.S.A.

Mathematics Department
Rice University
Houston, TX 77251
U.S.A.

D SUNDARARAMAN (d'après M KURANISHI)

Construction of the normal Cartan connection

I. Introduction

I.1. The Local Holomorphic Equivalence Problem

Let M_1, M_2 be two real analytic hypersurfaces in $\mathbb{C}^n$, $n \geqslant 2$. Take points $p_1 \in M_1$, $p_2 \in M_2$. M_1 and M_2 are said to be locally holomorphically equivalent at p_1 and p_2 if there exist open neighbourhoods V_1 of p_1 and V_2 of p_2 in $\mathbb{C}^n$ and a biholomorphic map $f: V_1 \longrightarrow V_2$ such that $f(V_1 \cap M_1) = V_2 \cap M_2$. Given a real analytic hypersurface M in $\mathbb{C}^n$, the local holomorphic equivalence problem for M is the problem of finding local differential invariants of M such that two such hypersurfaces are locally equivalent if and only if their corresponding invariants agree. This problem has been completely solved for any nondegenerate real analytic hypersurface M in $\mathbb{C}^n$, $n \geqslant 2$; by E. Cartan [1] in the two dimensional case and in higher dimensions first by Tanaka [6] and later independently by Chern and Moser [2] by a different method. Recently Kuranishi [5] has proposed another approach to the construction of the solution of the problem.

The aim of this paper is to describe the construction of Kuranishi and compare it with that of Chern-Moser. I am grateful to Professor Kuranishi for teaching me his theory; this paper is based largely on his preprint [4].

It is pertinent to recall here the relationship between the local holomorphic equivalence problem and the Riemann mapping problem. The Riemann mapping problem is the problem of deciding when two domains D_1, D_2 in $\mathbb{C}^n$, $n \geqslant 2$, are biholomorphically equivalent. A remarkable theorem of Fefferman [3] asserts that when D_1 and D_2 are bounded strictly pseudoconvex domains, the Riemann mapping problem is equivalent to the local holomorphic equivalence problem of their boundaries ∂D_1 and ∂D_2.

I.2. CR Structures And Their Defining Forms

Let M denote a differentiable (C^∞) manifold of dimension $(2n-1)$. By a CR structure of codimension 1 on M we mean a C^∞ subbundle $T''M$ of the complexified tangent bundle $\mathbb{C}TM$ of M such that

(1)
- i) $T''M$ has complex fibre dimension $(n-1)$
- ii) $T''M \cap T'M = \{0\}$ where $T'M = \overline{T''M}$

The CR structure $T''M$ is said to be *integrable* if the following integrability condition holds:

(2) iii) If $X, Y \in C^\infty(M, T''M)$, then their Lie bracket $[X, Y]$ also belongs to $C^\infty(M, T''M)$ where $C^\infty(M, T''M)$ denotes the space of C^∞ cross sections of the bundle $T''M$.

In this paper we consider only real analytic integrable CR structures of codimension 1. Such a CR structure is an abstraction of a real analytic real hypersurface M in $\mathbb{C}^n$: the CR structure of M is given by the subbundle $T''M$ of $\mathbb{C}TM$ consisting of the complex tangent vectors of type (0, 1). In this case, consider the subbundle $T^\circ M$ of TM defined by $T_p^\circ M = T_pM \cap JT_pM$ where J defines the complex structure of $\mathbb{C}^n$. It is easily seen that $T_p^\circ(M)$ is the maximal complex subspace of $T_p(M)$. The subbundle $T^\circ M = \bigcup_{p \in M} T_p^\circ$ is called the holomorphic tangent bundle of M. The CR structure of M is equivalently given by the pair $(T^\circ M, J)$. Given two such CR structures $(M_1, T^\circ M_1, J_1)$ and $(M_2, T^\circ M_2, J_2)$, a mapping $f: M_1 \longrightarrow M_2$ is called a CR mapping if and only if $df(T^\circ M_1) = T^\circ M_2$ and $df \circ J_1 = J_2 \circ df$ where d denotes the exterior derivative. A CR mapping f is called a CR equivalence if f^{-1} exists and is also a CR mapping. It is easy to check that the notions of CR equivalence and local holomorphic equivalence are the same.

Fix an integrable CR structure $^\circ T''M$ on M. For convenience we denote TM, $^\circ T''M$, $^\circ T'M$ simply by T, $T^{\circ\prime\prime}$, $^\circ T'$ respectively. Since $^\circ T' \oplus {}^\circ T''$ is preserved by conjugation there exists a subbundle T° of T of real codimension 1 whose complexification $\mathbb{C}T^\circ$ is $^\circ T' \oplus {}^\circ T''$. We can take a suitable real 1-form θ_0 on M such that the bundle T° is locally defined by $\theta_0 = 0$. Then we can choose 1-forms $\omega_0^1, \ldots, \omega_0^{n-1}$ on M such that $^\circ T''$ is defined locally by $\theta_0 = \omega_0^1 = \ldots = \omega_0^{n-1} = 0$ and $^\circ T'$ is defined locally by $\theta_0 = \overline{\omega_0^1} = \ldots = \overline{\omega_0^{n-1}} = 0$. Then, interpreting the defining conditions (i) and (iii) of $^\circ T''$, we get the following

Proposition. *A subbundle $^\circ T''$ of $\mathbb{C}TM$ is a CR structure if and only if $^\circ T''$ is locally defined by*

$$\theta_0 = \omega_0^1 = \ldots = \omega_0^{n-1} = 0, \qquad \theta = \overline{\theta} \tag{3}$$

for suitable Pfafiann forms $\theta_0, \omega_0^1, \ldots, \omega_0^{n-1}$ which satisfy the following conditions

$$\theta_0, \omega_0^1, \ldots, \omega_0^{n-1}, \overline{\omega_0^{n-1}}, \ldots, \overline{\omega_0^{n-1}}) \tag{4}$$

are linearly independent

$$\left.\begin{array}{l} d\theta_0 \equiv 0 \\ d\omega_0^j \equiv 0 \end{array}\right\} \text{ modulo } (\theta_0 . \omega_0^1, \ldots, \omega_0^{n-1}) \quad j = 1, \ldots, (n-1). \tag{5}$$

Remark: We refer to (θ_0, ω_0) as the defining forms of the CR structure $^\circ T''$.

I.3. The Levi Form of $^\circ T''$

Choose a real vector field S supplementary to T°. For $X, Y \in C^\infty(M, {}^\circ T'')$ define $L(X,Y)$ by $L(X, Y) \equiv L_S(X, Y)S \mod {}^\circ T' \oplus {}^\circ T''$ where

$$i[X, \overline{Y}] \equiv L_S(X, Y)S \qquad \mod {}^\circ T' \oplus {}^\circ T''. \tag{6}$$

$L(X, Y)$ is a Hermitian quadratic form defined on the fibres of $^\circ T''$. L may depend on the chosen vector field S. Since there is no intrinsic trivialization of the bundle $\mathbb{C}T/(°T'\oplus°T'')$, L is represented by the set $\{L_S\}$: they differ only by multiplicative functions. Hence L is well defined up to multiplication by nowhere vanishing real functions. In terms of the defining forms θ_0. $\omega_0=(\omega_0^1, ..., \omega_0^{n-1})$ we can write L, for a chosen S, as

$$L_S = \frac{1}{\theta_0(S)} C_{jk}\, \omega_0^j \otimes \omega_0^{\bar{k}}. \tag{7}$$

If L is nondegenerate, $°T''$ is said to be nondegenerate. We assume throughout that $°T''$ is nondegenerate. Consider a nondegenerate Hermitian form in $\mathbb{C}^{n-1}$ given by $\langle Z, Z\rangle = h_{j\bar{k}} Z^j \bar{Z}^k$ where $\overline{h_{j\bar{k}}} = h_{k\bar{j}}$ are constant. We assume that the CR structure $°T''$ is such its Levi form is similar to the above quadratic form, in the sense that at each point of M, there is an isomorphism i: $\mathbb{C}^{n-1} \longrightarrow T''$ such that $\langle Z, Z\rangle$ and $L_S\ \langle i(Z), i(Z)\rangle$ are proportional.

I.4. Cartan Connections

For solving the local equivalence problem for nondegenerate real analytic hypersurfaces M in $\mathbb{C}^2$, E. Cartan introduced a differential geometric structure (nowadays known as the normal Cartan connection) which solves the local holomorphic equivalence problem for M. He showed there exists a canonically constructed principal bundle $C \longrightarrow M$ (called the pseudoconformal bundle) with the normal Cartan connection such that two such hypersurfaces M_1 and M_2 are locally holomorphically equivalent if and only if the corresponding bundles $C_1 \longrightarrow M_1$, $C_2 \longrightarrow M_2$ are bundle equivalent with the corresponding normal Cartan connections being preserved. In higher dimensions also, Chern and Kuranishi have shown independently and by different methods that similar constructions can be employed. These approaches are analogous to the construction of connection and curvature on the orthonormal frame bundle of a Riemannian manifold giving a complete set of invariants for the Riemannian geometry.

Now we give the definition of a Cartan connection. Let G be a real Lie group and H be a closed subgroup of G. Let $\mathcal{G}, \mathcal{H}$ be respectively the Lie algebras of G and H. Let N be a differentiable manifold of dimension equal to that of G.

Definition. *(Cartan Connection). A Cartan connection for* N *with respect to* (G, H) *is a principal* H*-bundle* P $\longrightarrow$ N *together with a* $\mathcal{G}$*-valued Pfaffian form* ω *on* P *satisfying the following conditions:*

(8) *i)* $\forall\ x \in P$, ω *induces an isomorphism of* T_xP *with* $\mathcal{G}$.

ii) Identifying a fiber of P *with* H, ω *restricted to* H *must be the left invariant Maurer-Cartan form on* H.

iii) $R_h^*\omega = \mathrm{ad}(h^{-1})\omega$ *where* R_h *denotes the right action of* H *on* P *and* ad *represents the adjoint action of* H *on* $\mathcal{H}$.

For example, if $G \longrightarrow G/H$ is regarded as a principal H-bundle, then the Maurer-Cartan form on G defines a Cartan connection on the bundle, for any choice of H.

The solution to the local holomorphic equivalence problem for a nondegenerate CR structure on M is in essence, obtained by properly choosing (G, H) and constructing a unique Cartan connection in a canonically-constructed principal H-bundle $P \longrightarrow M$. In all the approaches, the choice of (G, H) is easily made but the constructions of the bundle $P \longrightarrow M$ with the normal Cartan connection are very difficult, as we shall see.

The Choice of (G, H): Let $(Z^1, \ldots, Z^{n-1}, w) \in \mathbb{C}^n$ where $w = x + iy$. Consider the nonsingular quadric Q in $\mathbb{C}^n$ defined by

$$y = \frac{1}{2}\sum_{j,k=1}^{n-1} h_{j\bar{k}} Z^j Z^{\bar{k}}$$

where $(h_{j\bar{k}})$ is a nonsingular Hermitian matrix. Q has the canonical nondegenerate real analytic CR structure $T''Q$ induced by $\mathbb{C}^n$. Consider the one point compactification $\tilde{Q}'$ of Q in $P''(\mathbb{C})$ See §6 of [5]. $T''Q$ extends to a unique CR structure $T''\tilde{Q}$ on $\tilde{Q}$. $\tilde{Q}$ is a homogeneous space: If (p, q) is the signature of $(h_{j\bar{k}})$, the group $G = \dfrac{SU(p+1, q+1)}{\mathbb{Z}_{n+1}}$ where $\mathbb{Z}_{n+1} = \{\lambda I, \lambda^{n+1} = 1\}$, acts transitively and effectively on $\tilde{Q}$. G is a group of CR automorphisms of $\tilde{Q}$. Let H be the isotropy subgroup of g at the origin. The map $G \longrightarrow \tilde{Q}$ defined by $g \longrightarrow g(0)$ induces an isomorphism of G/H onto Q. Hence it induces an isomorphism $G/H \longrightarrow T_0\tilde{Q} = T_0 Q$ and so $G/H \otimes \mathbb{C} \overset{\alpha}{\approx} \mathbb{C}T_0 Q$.

The above G, H turn out to be the correct choice for the construction of the normal Cartan connection.

Definition. *(Cartan Connection for a CR Structure): Let* $T''N$ *be a nondegenerate CR structure on a real manifold* N. *Let* (P, ω) *be a Cartan connection for* N *with respect to* (G, H). *We say* (P, ω) *is for the CR structure* $T''N$ *if for each* $x \in P$, $(\alpha \circ \text{projection} \circ \omega)^{-1}(T''Q)$ *projects into* $T''N$, *where*

$$\mathbb{C}T_x P \xrightarrow{\ \omega\ } G\otimes\mathbb{C} \xrightarrow[\text{projection}]{} (G/H)\otimes\mathbb{C} \xrightarrow{\ \overset{\alpha}{\approx}\ } \mathbb{C}T_0 Q.$$

II. Chern's Structure Bundle

We give a brief description of the Chern's structure bundle [2]. We follow closely the notes of Kuranishi ([5] these proceedings). We start with a fixed nondegenerate real analytic CR structure $^\circ T''$ on a real manifold M of dimension (2n–1). Let (θ_0, ω_0) be the Pfaffian forms defining $^\circ T''$.

Define:

(1) $E_1 = \{\theta \in T^*M | \theta$ is positive, θ defines the holomorphic tangent space T_x° at the source x of $\theta\}$

Then E_1 is a subbundle of T^*M. Let $\rho_1 \colon E_1 \longrightarrow M$ be the projection map. Define the tautology form Θ of E_1 over M by $(\Theta)_\theta = (\rho_1^* \theta)_\theta$ where the suffix θ denotes the point θ of E_1. Θ is a globally well defined Pfaffian form on E_1. Over an open set U, E_1 has a chart $U \times R^+ (x, p) \longrightarrow p(\theta_0)_x \in E_1$. Thus in terms of the chart (x, p), $\Theta = p\theta_0$.

Define:

(2) $E_2 = \{(\theta, \omega) | \theta \in E_1,\ \omega = (\omega^1, \ldots, \omega^{n-1})$ where $\omega^j \in T^*_x M,\ x = \rho_1 \theta$ such that:

(i) $\theta = \omega = 0$ defines T''_x

(ii) $(d\Theta)_\theta \equiv i\langle \rho_1^* \omega, \rho_1^* \omega \rangle \mod \Theta\}$

E_2 is a bundle over E_1 with projection ρ_2: $E_2 \longrightarrow E_1$ given by $\rho_2(\theta, \omega) = \theta$.

Define the tautology from $\Omega = (\Omega^1, \ldots, \Omega^{n-1})$ of E_2 over E_1 by

(3)
$$(\Omega^j)_{(\theta,\omega)} = (\rho_2^* \omega^j)_{(\theta,\omega)}.$$

Over an open set U, we can write for $(\theta, \omega) \in E_2$

(4)
$$\theta = p\theta_0, \qquad p > 0$$
$$\omega = \omega_0 t + v\theta \qquad \text{where} \qquad t = (t^i_j) \in GL(n-1, \mathbb{C}),$$
$$\text{and} \qquad v = (v^1, \ldots, v^{n-1}) \in \mathbb{C}^{n-1}.$$

The matrix t satisfies the condition $t^* t = pI$. (x, t, p, v) can be regarded as a chart for E_2 and v as the fiber chart of E_2 over E_1.

We have, in terms of the chart (t, p, v) of E_2 and (θ_0, ω_0),

(5)
$$\Omega = \omega_0 t + v\Theta$$

(6)
$$d\Theta = i\langle \Omega, \Omega \rangle + \Theta \wedge \pi$$

(7)
$$\pi = -d\log p + \pi_0 - i\langle v, \Omega \rangle + i\langle \Omega, v \rangle + s\Theta$$

where s is any real number and π_0 is the form given by

(8)
$$d\theta_0 = i\langle \omega_0, \omega_0 \rangle + \theta_0 \wedge \pi_0.$$

Define a bundle C over E_2 by

(9)

$$C = \{(\theta, \omega, \pi) | (\theta, \omega) \in E_2,\ \pi \in T^*_{(\theta,\omega)} E_2 \text{ such that } (d\Theta)_{(\theta,\omega)} = i\langle \Omega, \Omega \rangle_{(\theta,\omega)} + (d\Theta)_{(\theta,\omega)} \wedge \pi\}$$

Define the projection ρ_C: $C \longrightarrow E_2$ by

(10)
$$\rho_C(\theta, \omega, \pi) = (\theta, \omega)$$

Remark: In Kuranishi [5], the bundle C is denoted by E, and the map ρ_C by ρ_3. In Chern-Moser [2], the bundle is denoted by Y. For our convenience of comparison we have changed the notation.

We define the tautology form Π of C over E_2 as follows:

$$\Pi = -d\log p + \frac{1}{p}\pi_0 - i\langle v, \Omega\rangle + i\langle \Omega, v\rangle + s \tag{11}$$

We can regard (x, p, t, v, s) as a chart of C over an open set v and s as the fiber chart of C. The forms Θ, Ω, Π are invariently defined on C. Writing the expression for $d\Theta$, $d\Omega$, $d\Pi$ we see there exist locally defined forms τ, ϕ, ψ such that

$$\begin{aligned} d\Theta &= i\langle \Omega, \Omega\rangle + \Theta \wedge \Pi \\ d\Omega &= \Omega \wedge \tau + \Theta \wedge \phi \\ d\Pi &= i\langle \phi, \Omega\rangle + i\langle \Omega, \phi\rangle + \Theta \wedge \psi \end{aligned} \tag{12}$$

The forms ϕ, ψ are real and the complex matrix valued form τ satisfies $\tau + \tau^* = \Pi I$. They are not uniquely defined. It is easily seen that the globally invariantly defined forms Θ, Ω, Π together with the locally defined forms τ, ϕ, ψ form a local base of Pfaffian forms on C. In order to have a base by globally invariantly defined forms, we have to make the choices of τ, ϕ, ψ unique. This is done by calculating the curvature forms locally and then imposing suitable "normalizing" conditions on them so that τ, ϕ, ψ become unique. This is done in Chern-Moser [2]; also see Kuranishi [5] in these proceedings. The globally invariantly constructed forms, corresponding to the unique τ, ϕ, ψ are denoted by T, Φ, Ψ. T, Φ, Ψ can be expressed in terms of the chart of C. The forms Θ, Π, Ψ are real forms, Ω, Φ are $\mathbb{C}^{n-1}$-valued forms and T is a (n–1)×(n–1)-complex matrix-valued form with the condition $\mathrm{T} + \mathrm{T}^* = \Pi I$. The expression for $d\mathrm{T}, d\Phi, d\Psi$ and the "normalizing conditions" on the curvature forms can be explicitly written down. See Chern-Moser [2] or Section 5 in Kuranishi [5] for the explicit formulae.

It is shown in Chern-Moser [2] that the canonically constructed $C \longrightarrow M$ is a H-bundle and the canonically constructed forms Θ, Ω, Π. T, Φ, Ψ are the components of a unique *G*-valued 1-form on C defining the normal Cartan connection in $C \longrightarrow M$ for the given nondegenerate CR structure, with respect to (G, H). Moreover, the bundle $C \longrightarrow M$ and the forms $\Theta, \Omega, \Pi, \mathrm{T}, \Phi, \Psi$ are locally unique up to isomorphism (for the proof of the last statement see Section 5 of Kuranishi [5].)

III. Kuranishi's Structure Bundle

III.1: Kuranishi's construction is illustrated by pointing out the analogy in the case of a Riemannian manifold M. In the latter case, the normal Cartan connection is the Levi-Cività connection; G is the group of Euclidean motions of R^n and H is the isotropy group at the origin. First consider the flat Riemann manifold R^n. We identify R^n with G/H. The structure bundle $P \longrightarrow R^n$ is by definition taken to be the bundle $G \longrightarrow G/H$ where G/H is identified with the bundle of all orthonormal coframes of R^n. The Cartan connection ω is by definition, taken to be the left invariant Maurer-Cartan form on G. Thus in the flat case the normal Cartan connection (P_{R^n}, ω_{R^n}) is easily constructed. For a general Riemannian manifold M, we take the structure bundle $P_M \longrightarrow M$ to be the bundle of orthonormal coframes of M and we transplant the form ω_{R^n} at each point to this bundle as follows: for each $p \in M$, take an osculating chart $i; (M', p) \longrightarrow (R^n, 0)$ where M' is an open neighbourhood of p. By definition the osculation chart such that the Riemannian metric of M and the induced metric of M induced by ι, agree at p up to the second order derivatives at p. It is well known that such charts exist and provide the best possible approximation of M to R^n. Consider the bundles $(\iota_p)^*P_{R^n}$ and P_M; they are subbundles of the bundle of coframes of M. They agree at all points over p and have the same tangent space at each element of P_M over p. Hence we can transport the form ω_{R^n} by ι_p to P_M. Thus in the case of a Riemannian manifold the normal Cartan connection is constructed by using osculating charts.

Kuranishi adopts, in essence, the same approach for the construction of the natural Cartan connection for a nondegenerate real analytic CR structure by discovering the right notions corresponding to coframes, orthonormal coframes and osculating charts. However, the details in this case are considerably more involved as we shall see.

III.2: The Bundle of CR Hypercoframes of A CR Structure

Consider the quadric Q given by the equation (8) of §I.4. Consider the canonical CR structure $T''Q$ on Q. Then $T''Q$ is defined by Pfaffian forms θ_Q, $\omega_Q = (\omega^1_Q, \ldots, \omega^{n-1}_Q)$:

(1) $$T''Q\colon \theta_Q = 0 = \omega^1_Q = \ldots = \omega^{n-1}_Q$$

In terms of the charts (z, x),

(2)
$$\theta_Q = dx + \frac{\sqrt{-1}}{2} h_{j\bar{k}}(z^j dz^{\bar{k}} + z^{\bar{k}} dz^j)$$
$$\omega_Q = dz^k, \quad k = 1, \ldots, (n-1).$$

Hence

(2′)
$$d\theta_Q = i h_{j\bar{k}} \omega^j_Q \wedge \omega^{\bar{k}}_Q = i\langle \omega_Q, \omega_Q \rangle$$
$$d\omega^k_Q = 0$$

Set

(3) $$T'Q = \overline{T''Q}, \qquad {}^\circ TQ = (T'Q + T''Q) \cap TQ,\ SQ = TQ/{}^\circ TQ.$$

Then $\theta_Q = 0$ defines ${}^\circ TQ$. SQ is a line bundle over Q. Its dual bundle S^*Q is generated by θ_Q. For the CR structure ${}^{\circ}T''M$ on the manifold M, we have similarly defined line bundles SM and S*M. We have assumed that the Levi forms of ${}^{\circ}T''M$ and $T''Q$ are equivalent. Hence we can introduce orientations in SQ and SM consistent with the equivalence of the Levi forms of ${}^{\circ}T''M$ and $T''Q$. By choosing a positive direction in S*M, we declare that ${}^{\circ}T''M$ is positively oriented.

For any $p \in M$ let m_p denote the ideal of p: m_p is the set of all germs at p of complex valued functions vanishing at p. m_p is the maximal ideal in the ring of germs at p of such functions. For any C^∞ vector bundle $E \longrightarrow M$ and for any two ideals I_p, I'_p, $I_p \subset I'_p$ the following sequence is exact:

$$0 \longrightarrow I'_p E/I_p \longrightarrow E/I_p \longrightarrow E/I'_p \longrightarrow 0 \tag{4}$$

where

$$E/I_p = C^\infty(M, E)_p / I_p(C^\infty(M, E)_p)$$

(5)

$$I'_p E/I_p = (I'_p C^\infty(M, E)_p / I_p(C^\infty(M, E)_p)$$

In particular consider $I_p = m_p^2$, $I'_p = m_p$; we have the canonical identification of E/m_p with E_p and $m_p E/m_p^2$ with $T_p^* M \otimes E_p$. Then from the above exact sequence (4) we have the following exact sequence

$$0 \longrightarrow T_p^* M \otimes E_p \xrightarrow{\ \iota\ } E/m_p^2 \longrightarrow E_p \longrightarrow 0 \tag{6}$$

We note that there is no canonical splitting of the above exact sequences. However, using a local base of $C^\infty(M, E)_p$ induced by a basis of $C^\infty(M, E)$, we can split the above sequences. Then depending on the choice of the basis, we have

$$E/m_p^2 = E_p + \iota(T_p^* M \otimes E_p) \tag{7}$$

Definition. *(Hypercoframe): A hypercoframe for* M *is a pair* $(\underline{a}, \underline{b})$ *of linear maps where*

$$\underline{a}: T_0^* Q \longrightarrow T_p^* M \text{ is an isomorphism}$$

(8)

$$\underline{b}: S^* Q/m_0^2 \longrightarrow T^* M/m_p^2$$

such that we have the following commutative diagram:

$$\begin{array}{ccccccccc} 0 & \longrightarrow & T_p^* Q \otimes S_0^* Q & \longrightarrow & S^* Q/m_0^2 & \longrightarrow & S_0^* Q & \longrightarrow & 0 \\ & & \downarrow{\scriptstyle \underline{a} \otimes \underline{a}_0} & & \downarrow{\scriptstyle \underline{b}} & & \downarrow{\scriptstyle \underline{a}_0} & & \\ 0 & \longrightarrow & T_p^* M \otimes T_p^* M & \longrightarrow & T^* M/m_p^2 & \longrightarrow & T_p^* M & \longrightarrow & 0 \end{array} \tag{9}$$

where $a_0 = \underline{a}|S_0^*Q$ and the horizontal sequences are obtained as above (6). When (9) is satisfied we say $\underline{b}$ is over $(\underline{a}, a_0)$. Note that a coframe is defined only with respect to the manifold structure of M and as such is independent of the CR structure $°T''M$ of M. Using the structure $°T''M$, we introduce now charts on the set of all hypercoframes H_pM at p. Note $\underline{b}$ is determined by $\underline{b}(\theta_Q)_0$.

Let (θ_0, ω_0) define $°T''M$. Then $\theta_0, \omega_0, \ldots, \omega_0^{n-1}, \overline{\omega}_0^1, \ldots, \overline{\omega}^{(n-1)}$ from a base of $\mathbb{C}T^*M$. For $(\underline{a}, \underline{b}) \in H_pM$, set

$$(10) \qquad \begin{aligned} \underline{a}(\theta_Q)_0 &= a^0(\theta_0)_p + a^{0k}(\omega_0^k)_p + a^{0\overline{k}}(\omega_0^{\overline{k}})_p \\ \underline{a}(\omega_Q^\ell)_0 &= a^0(\theta_0)_p + a^{\ell k}(\omega_0^k)_p + a^{0\overline{k}}(\omega^k)_p \end{aligned}$$

where a^0 is real and the rest of the coefficients are complex numbers. With respect to the choice of the basis $(\theta_0, \omega_0^k, \omega_0^{\overline{k}})$ of $\mathbb{C}T^*M$, we can split the sequence (6) corresponding to $E = T^*M$. Then we can write

$$(11) \qquad \underline{b}(\theta_Q)_0 = a^0\{\theta_0\}_p + a^{0k}\{\omega_0^{\overline{k}}\}_p + a^{0\overline{k}}\,\omega_0^k{}_p + \iota\{b^0 \otimes \theta_p + b^k \otimes \omega_p^{\overline{k}} + b^{\overline{k}} \otimes \omega_p^k\}$$

where $b^0 \in T_p^*M$ and $b^k \in \mathbb{C}T_p^*M$, $k = 1, 2, \ldots, (n-1)$.

Conversely take $(a^0, a^\ell, a^{\ell k}, a^{\ell\overline{k}}, b^0, b^k)$ where a^0 is real, $a, a^{0k}, a^{\ell k}, a^{\ell\overline{k}}$ are all complex numbers, $b^0 \in T_p^*M$ and $b^k \in \mathbb{C}T_p^*M$. Further let $\underline{a}$ given by (10) be an isomorphism. Then we check easily that $(\underline{a}, \underline{b})$ where $\underline{b}$ is given by (11), is a unique hypercoframe. Thus $(a^0, a^\ell, a^{0k}, a^{\ell k}, a^{\ell\overline{k}}, b^0, b^k)$ provide a chart for H_pM, and $H_p(M)$ is a manifold. In terms of these charts, $HM = \bigcup_{p \in M} H_pM$ is a C^∞ bundle over M. This bundle is analogous to the bundle of coframes of a Riemannian manifold. Elements of H_pM are called the *hypercoframes* of the CR structure $°T''$ on M at p.

Consider the germ of a local diffeomorphism $f: (M, p) \longrightarrow (Q, 0)$ at p. Then f induces a hypercoframe $(\underline{a}, \underline{b})$ at p as follows. Set

$$(12) \qquad \begin{aligned} \underline{a} &= (df)^*/m_0 \in \mathrm{Hom}(T_0^*Q, T_p^*M) \\ \underline{b} &= ((df)^*/m_0^2)|(S^*Q/m_0^2) \in \mathrm{Hom}(S^*Q/m_0^2, T^*M/m_0^2) \end{aligned}$$

Then it is checked that $(\underline{a}, \underline{b})$ is a hypercoframe. We have the induced bundle map $[f]: HQ \longrightarrow HM$. If we write

$$(13) \qquad \begin{aligned} f^*\theta_Q &= F_0\theta_0 + F^{0k}\omega_0^{\overline{k}} + F^{0\overline{k}}\omega_0^k \\ f^*\omega_Q^\ell &= F^\ell\theta_0 + F^{\ell k}\omega_0^{\overline{k}} + F^{\ell\overline{k}}\omega_0^k \end{aligned}$$

where F^0 is a real function and the rest of the coefficients are complex valued functions on an open neighbourhood of p in M.

Define

$$a^0 = F^0(p), \quad a^{0k} = F^{0k}(p), \quad a^{\ell} = F^{\ell}(p), \quad a^{\ell k} = F^{\ell k}(p)$$

(14)

$$a^{\ell\bar{k}} = F^{\ell\bar{k}}(p), \quad \underline{b}(\theta_Q)_0 = \{f^*\theta_Q\}_p$$

Then f induces the hypercoframe $(\underline{a}, \underline{b})$ given by the above. We also get using (11),

(15) $$b^0 = (dF^0)_p, \quad b^k = (dF^{0k})_p$$

Definition. *(CR Hypercoframes): A hypercoframe* $(\underline{a}, \underline{b})$ *at* $p \in M$ *of a CR structure* $°T''$ *on* M *is called a CR hypercoframe at* p *if the following conditions are satisfied:*

(16)

(1*) $\underline{a}$ *sends* $(\mathbb{C}T_0Q/T''Q)^* \longrightarrow (\mathbb{C}T_pM/°T_p'')^*$

(2*) $\underline{a}$ *preserves the orientations*

(3*) *The image of* $\underline{b}$ *is in* S^*M/m_p^2

(4*) $(\underline{a}, \underline{b})$ *is induced by the germ of a local diffeomorphism* $f: (M, p) \longrightarrow (Q, 0)$

The set of all CR hypercoframes at p of the CR structure $°T''$ on M forms a C^∞ bundle over M. We denote this bundle by $K_M^{°T''} \longrightarrow M$. When the CR structure $°T''$ is understood, we denote this bundle simply by $K_M \longrightarrow M$ or $K \longrightarrow M$. $K_M \longrightarrow M$ is a subbundle of $HM \longrightarrow M$. With respect to the defining forms (θ_0, ω_0) of $°T''$ we introduced a chart $(a^0, a^{\ell}, a^{0k}, a^{\ell k}, a^{\ell\bar{k}}, b^0, b^k)$ for HM. In terms of this chart we want to interpret the above conditions (1) to (4). Let $(h_{j\bar{k}}^M)$ be the Levi form of $°T''$. Recall we have assumed that $(h_{j\bar{k}}^M)$ is equivalent to the Levi form $(h_{j\bar{k}})$ of the CR structure T_Q'' of the quadric Q in $\mathbb{C}^n$. We can write

(17) $$d\theta_0 = \begin{cases} i h_{j\bar{k}}^M \omega_0^j \wedge \omega_0^{\bar{k}} + (r^{\bar{k}}\omega_0^k + r^k \omega^{\bar{k}}) \wedge \theta_0 \\ \\ i\langle \omega_0, \omega_0 \rangle + \theta_0 \wedge \pi_0 \end{cases}$$

where r^k is a complex valued C^∞ function on M and $\pi_0 = -(r^{\bar{k}}\omega_0^k + r^k \omega_0^{\bar{k}})$.

Interpreting the conditions (1) to (4) of a CR hypercoframe at p in terms of the chart of HM, we get that (1) to (4) are respectively equivalent to the following

(1′) $a^{0k} = 0 = a^{k\ell}$

(2′) $a^0 > 0$

(18) (3′) $a^{0k} = 0 = b^k$

$$(4') \quad a^0 h^M_{jk}(p) = h_{\bar{s}\bar{t}} a^{s\bar{j}} \overline{a^{t\bar{k}}} + r^k(p) a^0 = 0$$

$$b^{0k} + \iota h_{s\bar{\ell}} a^s \overline{a^{\ell\bar{k}}} + r^k(p) a^0 = 0$$

where b^{0k} is given by

$$b^0 = b^{00}(\theta_0)_p + b^{0k}(\omega_0^{\bar{k}}) + b^{0\bar{k}}(\omega_0^k)_p$$

Hence we can take $(a^0, a^\ell, a^{\ell\bar{k}}, b^0)$ as the chart for $K \longrightarrow M$ subject to the conditions (4′).

This bundle $K \longrightarrow M$ is analogous to the bundle of orthonormal coframes of a Riemannian manifold. The normal Cartan connection that we are after is going to be constructed in this bundle. Hence $K \longrightarrow M$ corresponds to the Chern's structure bundle $C \longrightarrow M$ discussed in Section II.

III.3. Normal Admissible Maps

As we remarked in the introduction of this section, the normal Cartan connection (the Levi-Cività connection) is constructed by using osculating charts. The correct analogue for osculating charts in Kuranishi's theory are the "normal admissible maps". However, the existence of normal admissible maps is a difficult existence theorem of the theory.

Definition. *(Admissible Maps): A local diffeomorphism* $h: (Q, 0) \longrightarrow (M, p)$ *is called admissible if the induced bundle map* $[h]: HM \longrightarrow HQ$ *satisfies the following conditions*

$$1°) \quad [h](K^{\circ T''M}) = K^{T''Q}$$

$$(19) \qquad 2°) \quad [h](K^{\circ T''M})_p = (K^{T''Q})_0$$

$$3°) \quad T_x([h](K^{\circ T''})) = T_x(K^{T''Q}) \quad \textit{for all } x \in K^{T''Q} \textit{ over } 0 \in Q.$$

In terms of the defining forms (θ_0, ω_0) of $°T''M$ and (θ_Q, ω_Q) of $T''Q$, we give a characterisation of admissible maps. For a local diffeomorphism $f: (Q, 0) \longrightarrow (M, p)$ we can write

$$h^*\theta_0 = H^{00}\theta_Q + H^{0\bar{j}}\omega_Q^j + H^{0j}\omega_Q^{\bar{j}}$$

(20)

$$h^*\omega_0^k = H^{k0}\theta_Q + H^{k\bar{j}}\omega_Q^j + H^{kj}\omega_Q^{\bar{j}}$$

Where $H^{\alpha\beta}$ are C^∞ functions on Q such that $\overline{H^{\alpha\beta}} = H^{\bar{\alpha}\beta}, \alpha \neq 0, \beta \neq 0$. Then we can prove the following proposition.

Proposition. *(Characterisation of Admissible Maps). A local diffeomorphism* $f: (Q,0) \longrightarrow (M,p)$ *is admissible if and only if*

$$1°) \quad H^{00}(0) > 0$$

$$(21) \qquad 2°) \quad H^{0k} \equiv 0 \mod m_0^3$$

3°) $H^{jk} \equiv 0 \mod m_0^2$

Then using the above characterisation we have the following important result whose proof is quite involved.

Proposition. *(Existence of Admissible Maps). Let* $°T''M$ *be a nondegenerate real analytic CR structure on a real manifold* M *with* $\dim M = (2n-1)$. *Let* Q *be the nonsingular quadric in* $\mathbb{C}^n$ *with its standard pseudoconvex CR structure* $T''Q$. *Then at any point* $p \in M$ *there exist admissible local diffeomorphisms* $h: (Q,0) \longrightarrow (M,p)$.

Using admissible local maps, it is possible to construct Cartan connections in $K \longrightarrow M$. However they turn out to be nonunique. With approximate normalisation conditions on the admissible maps, Kuranishi shows that we can construct a unique Cartan connection. This situation is analogous to the corresponding situation in Chern's theory. To state the normalisation conditions, we need the notion of weight of a function defined on Q. Choose global charts $(z, x), z = (z^1, \ldots, z^{n-1}),\ x = \mathrm{Re}\, Z^n$. Consider a polynomial function $F(Z, \overline{Z}, x)$. We assign weight to $F(Z, \overline{Z}, x)$ by assigning the weight 2 to x and 1 to $(Z, \overline{Z})$.

Any C^∞ function on Q can be written as a formal sum of homogeneous functions in $(Z, \overline{Z})$, coefficients being C^∞ functions of x. We can decompose the homogeneous part into type (ℓ, m) in $Z^j, Z^{\bar{j}}$. Thus we can speak of the type (ℓ, m) of the function.

Define the operator F by

$$(22) \qquad F = h^{j\bar{k}} \frac{\partial^2 F}{\partial Z^j \partial Z^{\bar{k}}} \quad \text{where} \quad h^{j\bar{k}} = (h_{j\bar{k}})^{-1}$$

and set

$$(23) \qquad Z^{\bar{k}} = \partial/\partial Z^{\bar{k}} - (i/2) h_{j\bar{k}} Z^j (\partial/\partial x)$$

Definition. *(Normal Admissible Maps). An admissible local diffeomorphism* $f: (Q,0) \longrightarrow (M,p)$ *is called normal at* p *if the following conditions are satisfied:*

(24)

(N_0) *The weight 3 part of* H^{0k} *is of the type* $(2,1)$

(N_1) $H^{\alpha k} \equiv 0, \sum_k Z^{\bar{k}} H^{\alpha \bar{k}} \equiv 0 \mod m_0^3,\ \alpha = 0, 1, \ldots, (n-1)$.

(N_2) $\Box(H^{\alpha k}) \equiv 0, \Box(\sum_k Z^{\bar{k}} H^{\alpha \bar{k}}) \equiv 0 \mod m_0^2,\ \alpha = 0, \ldots, (n-1)$

(N_3) $\Box^2 (H^{0k}) \equiv 0, \Box^2(\sum_k Z^{\bar{k}} H^{0\bar{k}}) \equiv 0 \mod m_0$.

Remark: The above conditions are analogous to the Chern's normalisation conditions on the curvature. The following properties can be checked.

(25) (a) If a particular choice of defining forms (θ_0, ω_0) of $°T''$ satisfies the above normalisation conditions, then any other choice of defining forms (θ, ω) of $°T''$ also satisfies the above conditions.

(26) (b) If $f: (Q,0) \longrightarrow (M, p)$ is normal at p, then $f \circ h$ is also normal at p for any element $h \in H$.

(27) (c) If a normal map $f: (Q,0) \longrightarrow (M,p)$ exists it is unique up to decomposition by elements of H; unique in the following sense:

If f_1, f_2 are normal at p, there is $g \in H$ such that

(28)
$$Z \circ f_1 - Z \circ f_2 \circ g \equiv 0 \quad \mathrm{mod}\, m_0^3$$
$$x \circ f_1 - Z \circ f_2 \circ g \equiv 0 \quad \mathrm{mod}\, m_0^4$$

Next we state one of the main theorems of Kuranishi's theory.

Theorem. *(Existence of Unique Normal Admissible Maps). Consider a nondegenerate real analytic CR structure* $°T''$ *on a real manifold* M, $\dim M = (2n-1)$*. Let* Q *be the nonsingular quadric in* $\mathbb{C}^n$ *with its standard strictly pseudoconvex CR structure* $T''Q$*. Then at any point* $p \in M$*, there exists a unique normal map* $f: (Q,0) \longrightarrow (M,p)$.

III.4: The Normal Cartan Connection for $T''Q$ in $K_Q^{T''Q} \longrightarrow Q$ With Respect to (G, H)

We indicate in this section how to construct the normal Cartan connection for the standard CR structure $T''Q$ on the quadric Q in $\mathbb{C}^n$. As mentioned earlier G is taken to be the group of local automorphisms of the CR structure $T''Q$. So any $g \in G$ is represented by a CR diffeomorphism $g: (Q', 0) \longrightarrow (Q'', g(0))$ where $Q'(Q'')$ is an open neighbourhood of 0 (of g(0)). The isotropy group of the operations of G on Q at 0 is taken to be the group H. It will be clear in the sequel that the structure group of the bundle $K_Q^{T''Q} \longrightarrow Q$ is H. The normal Cartan connection for the CR structure $T''Q$ will be constructed in this bundle. We denote $K_Q^{T''Q}$ by K_Q.

Take $g \in G$. Let $g^{-1}_{g(0)}$ be the CR hypercoframe induced by g^{-1} at g(0). Thus we get a C^∞ map

(29)
$$\iota_G : G \longrightarrow K_Q$$

Choose (θ_Q, ω_Q) which define $T''Q$ we can write

(30)
$$(g^{-1})^* \theta_Q = p_g \theta_Q$$
$$(g^{-1})^* \omega_Q^k = t_g^{k\bar{j}} (g_g^j \theta_Q + \omega_Q^j)$$

where $p_g, t_g^{kj}, t_g^{k\bar{j}}, q_g^j$ are C^∞ functions on the neighbourhood Q'' of g(0). Note that $(p_g(g(0)), q_g^j(g(0)),$

$t_g^{kj}(g(0)))$ give the fiber chart of $\iota_G(g)$. It can be shown that functions together with $\partial/\partial x(\log p_g)$ satisfy a completely integrable system of first order partial differential equations. It will then follow, by Frobenius' theorem, that the map ι_G is injective. Then it is shown that ι_G is bijective. Hence ι_G is a diffeomorphism and G is the global automorphism group. Further we check that $\iota_G(g_1 g_2) = g_1^{-1}\iota_G(g_2)$. For later reference we include these results in the following proposition.

Proposition. *The map* $\iota_G\colon G \longrightarrow K_Q$ *has the following properties:*

(31)
(i) ι_G *is a* C^∞ *diffeomorphism*

(ii) $\iota_G(g_1 g_2) = g_1^{-1}\,\iota_G(g_2)$ *for any* $g_1, g_2 \in G$.

We define functions p_G, q_G, t_G, p'_G on G by

(32)
$$p_G(g) = p_g(g(0)), \qquad q_G(g) = q_g(g(0))$$
$$t_G(g) = t_g(g(0)), \qquad p'_G(g) = p'_g(g(0)$$

Define the map $\pi\colon G \longrightarrow Q$ by $\pi(g) = g(0)$. In terms of the defining forms (θ_Q, ω_Q) of $T''Q$ and the global coordinates (z, x) of Q, we can express the charts of K_Q, G and H as follows:

(33) K_Q has chart $(Z, x, a^0, a^k, a^{k\bar{\ell}}, b^0)$

(34) G has a chart $(z\circ\pi, x\circ\pi, p_G, V_G, t_G, p_G p'_G)$ where $V_G = t_G q_G$, $t_G^* t_G = p_G I$ where $*$ denotes the adjoint with respect to the pairing $\langle\ ,\rangle$.

(35) H has chart (p_G, V_G, t_G, p'_G) where $t_G t_G^* = p_G I$.

The map ι_G is given under the above charts, by

(36) $$z = z\circ\pi, \quad x = x\circ\pi, \quad a^0 = p_G, \quad a^k = V_G^k, \quad a^{k\bar{\ell}} = (t_G)^k_{\bar{\ell}}, \quad b^0 = p_G p'_G$$

By means of the diffeomorphism ι_G, Pfaffian forms on G can be transported to K_Q. By transporting to K_Q suitable canonically constructed forms on G, we can construct the normal Cartan connection in $K_Q \longrightarrow Q$. The required forms on G, are defined as follows. Define the left invariant forms

(37) $$\Theta_G,\ \Omega_G,\ \Pi_G,\ \Phi_G,\ T_G,\ \Psi_G$$

as those which agree at the identity with, respectively,

(38) $$\pi^*\theta_Q,\quad \pi^*\omega_Q,\quad d\log p_G,\quad dq_G,\quad dt_G,\quad dp'_G$$

where x denotes the adjoint with respect to the pairing (2′).

The relation $t_G t_G^* = p_G I$ leads to

(39) $$\tau_G + T_G^* = \Pi_G I$$

Now the required forms on K_Q are

(40) $$\iota_G^*(\Theta_G),\quad \iota_G^*(\Omega_G),\quad \iota_G^*(\Pi_G),\quad \iota_G^*(\Phi_G),\quad \iota_G^*(T_G),\quad \iota_G^*(\Psi_G)$$

These define the normal Cartan connection in $K_Q \longrightarrow Q$. To see that this is a H-bundle, recall that H has chart (p_G, V_G, t_G, p_G') where $T_G T_G^* = p_G I$. If $g \in G$, $h \in H$ then we check the following

(41) $$\begin{aligned} p_G(gh) &= p_G(g)\, p_G(h) \\ q_G(gh) &= q_G(g) + p_G(g)\, t_G(g) t_G(g)^{-1} q_G(h) \\ t_G(gh) &= t_G(h) t_G(g) \\ p_G'(gh) &= p_G'(g) + p_G(g) p_G'(h) + 2R i\langle t_G(g)\, q_G(g), q_G(h)\rangle \end{aligned}$$

where IR denotes real part.

(42) The required right action R_h of H is obtained by defining $R_h(g) = gh$.

In terms of the defining forms (θ_Q, ω_Q) of $T''Q$ and the chart of G, we can explicitly write down the expressions for the canonical forms $\Theta_G, \Omega_G, \Pi_G, \Phi_G, T_G, \Psi_G$, for $d\Theta_G, d\Omega_G, d\Pi_G, d\Phi_G, dT_G, d\Psi_G$ and also for $R_h^*\Theta_G$, $R_h^*\Pi_G$, $R_h^*\Pi_G$, $R_h^*\ {}^*_G$, $R_h^* T_G$, and $R_h^*\Psi_G$. We state these below:

(i) $\Theta_G = p_G \theta_Q$

(ii) $\Omega_G = t_G \omega_Q + v_G \theta_Q$ where $v_G = t_G q_G$

(iii) $\Pi_G = d\log p_G - \frac{1}{p_G} p_G' \Theta_G - 2R \frac{i}{p_G} \langle \Omega_G, v_G \rangle$

(43) (iv) $T_G u = (dt_G) t_G^{-1} u - (\frac{1}{2p_G} (p_G' - \frac{i}{p_G}\langle v_G, v_G\rangle) u - \frac{i}{p_G^2}\langle u, v_G\rangle \Theta_G - \frac{i}{p_G}\langle \Omega_G, v_G\rangle u$

$$- \frac{i}{p_G}\langle u, v_G\rangle \Omega_G - \frac{i}{p_G}\langle u, \Omega_G\rangle v_G$$

(v) $\Phi_G = \frac{1}{p_G} dv_G - \frac{1}{p_G} T_G v_G - \frac{1}{2p_G}(p_G' - \langle v_G, v_G\rangle)(\Omega_G + \frac{1}{p_G} v_G \Theta_G) - \frac{i}{p_G^2}(\langle \Omega_G, v_G\rangle v_G + \langle v_G, v_G\rangle \Omega_G)$

(vi) $\Psi_G = \frac{1}{p_G} dp_G' + 2R\frac{i}{p_G}\langle \Phi_G, v_G\rangle - \frac{1}{2p_G^2}(p_G'^2 + \frac{1}{p_G^2}\langle v_G, v_G\rangle^2)\Theta_G - R\frac{i}{p_G^2}(p_G' + \frac{i}{p_G}\langle v_G, v_G\rangle)\langle \Omega_G, v_G\rangle$

(44)

(i) $d\Theta_G = i\langle \Omega_G, \Omega_G\rangle + \Pi_G \wedge \Theta_G$

(ii) $d\Omega_G = T_G \wedge \Omega_G + \Phi_G \wedge \Theta_G$

(iii) $d\Pi_G = i\langle\Omega_G, \Phi_G\rangle + i\langle\Phi_G, \Omega_G\rangle + \Theta_G \wedge \Psi_G$

(iv) $d\Phi_G = T_G \wedge \Phi_G - \Pi_G \wedge \Phi_G + \frac{1}{2}\Omega_G \wedge \Psi_G$

(v) $dT_G u = -T_G \wedge T_G u + \frac{1}{2}\Theta_G \wedge T_G u - i\Omega_G \wedge \langle u, \Phi_G\rangle + i\langle u, \Omega_G\rangle \wedge \Phi_G + i\langle\Omega_G, \Phi_G\rangle u$

(vi) $d\Psi_G = \Psi_G \wedge \Pi_G - 2Ri\langle\Phi_G, \Phi_G\rangle$

Remark: The above are calculated at the identity (as it is sufficient because of the left invariance of the forms).

(i) $R_h^*\Theta_G = p_G(h)\,\Theta_{G*}$

(ii) $R_h^*\Omega_G = t_G(h)\Omega_G + t_G(h)q_G(h)\Theta_G$

(iii) $R_h^*\Pi_G = \Pi_G - p_G'(h)\Theta_G - 2Ri\langle\Omega_G, q_G(h)\rangle$

(45)

(iv) $R_h^*\Phi_G = \frac{1}{p_G(h)} t_G(h)(\Phi_G + q_G(h)\Pi_G - T_G q_G(h)) - \frac{1}{2p_G(h)}(p_G'(h)$

$- i\langle q_G(h), q_G(h)\rangle) t_G(h)(\Omega_G + q_G(h)_G) + \frac{i}{p_G(h)}\langle q_G(h), \Omega_G\rangle t_G(h) q_G(h),$

(v) $R_h^* T_G u = t_G(h) T_G t_G(h)^{-1} u - (\frac{1}{2}(p_G'(h) + i\langle q_G(h), q_G(h)\rangle) u + \frac{i}{p_G(h)}\langle u, t_G(h)q_G(h)\rangle t_G(h) q_G(h))\Theta_G$

$-i\langle\Omega_G, q_G(h)\rangle u - \frac{i}{p_G(h)}\langle u, t_G(h)q_G(h)\rangle t_G(h)\Omega_G - \frac{i}{p_G(h)}\langle u, t_G(h)\Omega_G\rangle t_G(h) q_G(h),$

(iv) $R_h^*\Psi_G = \frac{1}{p_G(h)}(\Psi_G + p_G'(h)\Pi_G + 4Ri\langle\Phi_G, q_G(h)\rangle - 2Ri\langle T_G q_G(h), q_G(h)\rangle) - \frac{1}{2p_G(h)}(p_G'(h)^2$

$+ \langle q_G(h), q_G(h)\rangle)\Theta_G - 2R\frac{i}{p_G(h)}(p_G'(h) - i\langle q_G(h), q_G(h)\rangle)\langle\Omega_G, q_G(h)\rangle$

Remark: It is sufficient to check that the above are valid at the identity element.

III.5: The Normal Cartan Connection for $^\circ T''M$ **in** $K_M^{\overset{\circ}{T}''M} \longrightarrow M$ **With Respect to** (G, H)

We denote $K_M^{\overset{\circ}{T}''M}$ simply by K. We claim that $K \longrightarrow M$ is a H-bundle. We define the action of H on K as follows: Take $X \in K$. Then X is a CR hypercoframe at $p \in M$ induced by a germ of local diffeomorphism f: (m, p) $\longrightarrow$ (O, 0). Let $h \in H$. We define $R_h X$ to be the CR hypercoframe at $p \in M$, represented by $h^{-1} \circ f$. Consider the diffeomorphism $\iota_G: G \longrightarrow K$ (see (1) of III.4). Then $[f]\iota_G(I)$ is the hypercoframe at p represented by f, where I is the identity of G. The following relations are easily verified:

(46)

$$R_h([f]\iota_G(I)) = [h^{-1} \circ f]\iota_G(I) = [f](\iota_G(h_1 h))$$

$$R_h([f]\iota_G(h_1)) = [f](\iota_G(h_1 h))$$

Hence under the fiber chart of H, $h_1 \longrightarrow [f](\iota_G(h_1))$ and thus R_h is represented by the right multiplication by h.

We now define the required connection form ω: $T(K) \longrightarrow g$ as follows. Fix $p \in M$. Consider a germ of normal admissible map h_p: $(Q, 0) \longrightarrow (M, p)$ (refer the theorem of III.3). Thus for each $x \in K$ over p, we have

(47) $$[dh_p]_X: T_X(K) \longrightarrow T_{[h_p]X}(K^{T''Q})$$

By the diffeomorphism ι_G: $G \longrightarrow K^{T''Q}$, $T_{[h_p]X}(K^{T''Q})$ can be identified with $T_I G$, by sending $g \longrightarrow ([h_p]x)^{-1}g$, $g \in G$. Since $T_I G$ is the Lie algebra g of G, we have a map

(48) $$[d\,h_p]_X: T_X(K) \longrightarrow g$$

It can be shown that $[d\,h_p]_X$ is independent of the h_p chosen. We define

(49) ω: $T(K) \longrightarrow g$ by $\omega|T_X(K) = [d\,h_p]_X$ for all $p \in M$ and X above p.

It is checked from our construction that ω is the sought after unique Cartan connection for $°T''M$ in $K \longrightarrow M$. Finally we arrive at the main theorem of Kuranishi:

Theorem. *(Existence of Unique Normal Cartan Connection for* $°T''M$*). Let* $°T''M$ *be a real analytic nondegenerate CR structure on the real manifold* M, $\dim M = (2n-1)$*. Then there exists a canonically constructed* H*-bundle* $K_M^{°T''M} \longrightarrow M$ *and a unique Cartan connection* ω: $T(K_M^{°T''M}) \longrightarrow g$ *solving the local equivalence problem for* $°T''M$.

The canonical forms determining the normal Cartan connection, denoted by $\Theta_{°T''}, \Omega_{°T''}, \Pi_{°T''}, \Phi_{°T''}, T_{°T''}, \Psi_{°T''}$, are defined by

(50) $$(\Theta_{°T''})_X = [h_p]^*(\Theta_G)_{[h_p]X}$$

and similarly for the remainig forms.

The canonical forms can be explicitly written in terms of charts of G and K_M.

IV. Identification of the Kuranishi Structure Bundle with the Chern Structure Bundle

Let $°T''$ be a nondegenerate real analytic CR structure of codimension 1, on a real manifold M, $\dim M = 2n-1$. We denote the Chern structure bundle for $°T''$ by $C \longrightarrow M$ and the Kuranishi structure bundle for $°T''$ by $K \longrightarrow M$. The normal Cartan connection for $°T''$ in $C \longrightarrow M$ with respect to (G, H) is determined by the canonically constructed forms $\Theta_C, \Omega_C, \Pi_C, T_C, \Phi_C, \Psi_C$ (refer Section II and for full account see Kuranishi [5] in these proceedings). The normal Cartan connection for $°T''$ in $K \longrightarrow M$ with respect to (G, H) is determined by the canonically constructed forms Θ_C, $\Omega_K, \Pi_K, T_K, \Phi_K, \Psi_K$ (refer Section III). In this section we show that these two bundles $C \longrightarrow M$

and $K \longrightarrow M$ are isomorphic with the corresponding normal Cartan connections being identified:

Theorem. *Let* $°T''$ *be a nondegenerate real analytic CR structure on a real manifold* M. *Then its Chern structure bundle* $C \longrightarrow M$ *and the Kuranishi structure bundle* $K \longrightarrow M$ *are isomorphic with the corresponding normal Cartan connection being preserved.*

Let (θ_0, ω_0) define $°T''$. Then with respect to (θ_0, ω_0), K has fiber chart $(a^0, a^k, a^{k\bar{\ell}}, b^0)$ (see equation (33) of Section III). E_2 has fiber chart (p, t, v) (see equation (4) of section II).

We define

(1) $$\rho_K: K \longrightarrow E_2 \quad \text{by} \quad \rho_K(\underline{a}, \underline{b}) = (\theta, \omega)$$

where $\theta = \underline{a}((\theta_Q)_0)$, $\omega^k = \underline{a}((\omega^k_Q)_0)$ and $\omega = (\omega^1, \ldots, \omega^{n-1})$. Under this map the charts of K and C are related as follows:

(2) $$a^0 = p, \quad a^k = \vartheta p, \quad a^{k\bar{\ell}} = t^k_\ell.$$

Thus we have the following situation

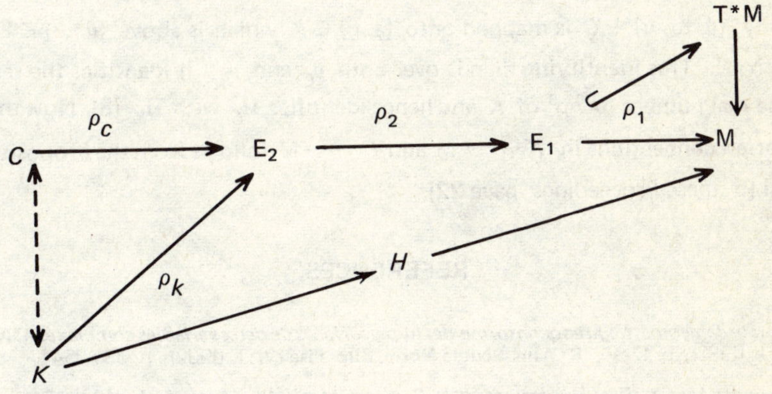

where H denotes the bundle of hypercoframes on M.

Let $(\underline{a}, \underline{b}) \in K$ at $p \in M$ and $\rho_K(\underline{a}, \underline{b}) = (\theta, \omega) \in E_2$. We denote $(\theta, \omega) \in E_2$ by P. Let $\rho = \rho_1 \circ \rho_2$. We have the induced map ρ_*: $\mathbb{C}TE_2 \longrightarrow \mathbb{C}TM$. Now $°T'' + °T'$ is of codimension 1 in $\mathbb{C}TM$. Denote $\rho_*^{-1}(°T'' + °T')$ by $T°E_2$. Let $\mathbb{C}TE_2/T°E_2$ be denoted by SE_2. The projection ρ: $E_2 \longrightarrow M$ induces an injective map

(4) $$\rho^*_{p,P}: S^*_p M \longrightarrow S^*_p E_2.$$

We can prove the following relations:

(5) $$\rho^*_{p,P}(b(\theta_Q \mod m_0^2)) = (\Theta_K)_P \mod m_P^2 + \iota((\Theta_K)_P \otimes \beta)$$

where $\beta \in \mathbb{C}T^*_p E_2$ is given by

(6) $$\beta = \rho^*_{p,P}(-d\log p) + \rho^*_{p,P}(\frac{1}{p}(dF^\circ)_P)$$

where p is the fiber coordinate of E_1 (see Section II) and F° is the real valued function (given in equation (13) of Section III) defined in a neighbourhood of p and ι is given in (7) of Section III. Note that for the above β, we have

(7) $$(d\Theta_C)_P = i\langle(\Omega_C)_P, (\Omega_C)_P\rangle + \Theta_C \wedge \beta$$

This is clear from equation (6) of Section II.

(8) $$(\Pi_K)_P = -d\log a_0 + \pi_0 + 2\mathrm{IRi}\langle(\Omega_K)_P, v\rangle + \frac{b}{p^2}(\Theta_K)_P.$$

The proof of the theorem follows from above (5) and (8). Take $(\underline{a}, \underline{b}) \in K$ above $(\Theta, \omega) \in E_2$. We map $(\underline{a}, \underline{b})$ to $(\theta, \omega, \beta) \in C_2$. The expression (6) for ρ shows that in fact $(\theta, \omega, \beta) \in C$. On the other hand any $(\theta, \omega, \pi) \in C$ is mapped onto $(\underline{a}, \underline{b}) \in K$ which is above $(\theta, \omega) \in E_2$. Thus C is identified with K. This identifying map is over both ρ_C and ρ_K; it identifies the real number s of C with the real number b^{00}/p^2 of K and hence identifies Π_C with Π_K (8). Now the identity of the normal Cartan connections in $K \longrightarrow M$ and $C \longrightarrow M$ follows from the Proposition in Section 5 of Kuranishi [5, these Proceedings, page 22].

REFERENCES

[1]. E. Cartan, *Sur la geometrie psedoconforme des hypersurfaces de deux variables complexes*, I: **Ann. Math. Pure Appl.** (4) II (1932), p. 17-90; II; **Ann. Scuola Norm. Sup. Pisa** (2) I (1932), p. 333-354.;

[2]. S.S. Chern and J. Moser, *Real Hypersurfaces in Complex Manifolds,* **Acta Math.** 113 (1974), p. 219-271.

[3]. C. Fefferman: *The Bergman Kernel and Biholomorphic Mappings of Pseudoconvex Domains,* **Invent. Math.** 26 (1974), p. 1-65.

[4]. M. Kuranishi: *The Normal Cartan Connection on a CR Structure,* Preprint.

[5]. M. Kuranishi: *Local Geometry of Nondegenerate CR Structures,* These Proceedings.

[6]. N. Tanaka: *On the Pseudoconformal Geometry of Hypersurfaces of the Space of n Complex Variables,* **J. Math. Soc. Japan** 14 (1962), p. 397-429; *On The Nondegenerate Real Hypersurfaces, Graded Lie Algebras and Cartan Connections,* **Japanese J. Math.** Vol. 2 No. 1 (1976), p. 131-190.

Centro de Investigación y de Estudios Avanzados del IPN
Apartado Postal 14-740
México 14, D.F.
México

L M TOVAR

Open Stein subsets and domains of holomorphy in complex spaces

If A_1 and A_2 are open Stein subsets in a complex space X, it is well known that $A_1 \cup A_2$ is not necessarily Stein. Another important problem is the following: let $\Omega_1 \subset \Omega_2 \subset \ldots$ be a sequence of open Stein subsets in X and let $\Omega = \bigcup_{j=1}^{\infty} \Omega_j$. If X is a Stein manifold, it is known that Ω is Stein. If X is a Stein space, it is not known whether Ω should be Stein. J. E. Fornaess [11] has given an example of a sequence of increasing Stein subsets in a manifold whose union is not Stein. Thus we have the following famous problem: Let X be a complex space and $\Omega_1 \subset \Omega_2 \subset \ldots$ be a sequence of open Stein subsets in X and let $\Omega = \bigcup_{j=1}^{\infty} \Omega_j$. The question is, is Ω Stein?

Theorems 1, 3 and 4 of this paper contain some results related to these problems. Theorem 2 characterizes open Stein subsets in terms of the Cousin I property. Since the proof of Theorem 1 is very long, I give only an outline of the proof. The proofs of theorems 2, 3, 4 are based on the techniques of E. Ballico [10].

I would like to express my thanks to my advisor Dr. E. Ramírez de Arellano, who revised and presented these results at the "Taller de Varias Variables Complejas" in August, 1983, in "La Trinidad", Tlaxcala, México, while I was out of the country. Also I am very grateful to Dr. Hugo Rossi for his comments and suggestions.

In this paper X will denote a holomorphically separable complex space of finite dimension. For the basic notions and properties of Stein spaces we refer to [1].

Theorem 1. *Let* X *be a normal Stein space,* $\Omega_1 \subset \Omega_2 \subset \ldots$ *a sequence of open Stein subsets of* X *and let* $\Omega = \bigcup_{j=1}^{\infty} \Omega_j$ *be irreducible. Let* S *be the singular set of* X. *If* $\Omega \subset\subset X$ *and* $\mathrm{Fr}\,\Omega \backslash S$ *is dense in* $\mathrm{Fr}\,\Omega$, *then* Ω *is a domain of holomorphy.*

Outline of proof:

I. For every $p \in (\mathrm{Fr}\,\Omega \backslash S)$ and every sequence $z_n \longrightarrow p$ in Ω, we show the existence of a holomorphic function in Ω which diverges in $\{z_n\}$. The argument takes the following steps:

a) We can find holomorphic mappings $\Phi_r: X \longrightarrow \mathbb{C}^n$, $r = 1, 2, \ldots n$ such that:

i) Each Φ_r has discrete fibers

ii) $S = \bigcap_{r=1}^{\ell} Z'_r$, where Z'_r is the ramification locus of X with respecto to Φ_r.

Further, there exist holomorphic functions $f_1, f_2, \ldots f_\ell$ over X such that if $Z_r = \{x \in X: f_r(x) = 0\}$, $r = 1, 2, \ldots \ell$, then $Z'_r \subset Z_r$ and $S = \bigcap_{r=1}^{\ell} Z_r$ [see [5]]. Then consider the unramified domain

$$\Phi_r: \Omega\backslash Z_r \longrightarrow \mathbb{C}^n.$$

Observe that each $(\Omega\backslash Z_r)$ is Stein. By using the techniques of references [3] and [6] we construct continuous plurisubharmonic functions φ_r on Ω which are zero in Z_r and diverge to ∞ when we go to $(Fr\Omega\backslash Z_r)$.

b) Take $p \in (Fr\Omega\backslash S)$ and choose "r" such that $p \notin Z_r$. Then it is possible to find holomorphic functions $h_1, h_2, \ldots h_m$ in X with $p = \{x \in X: h_i(x) = 0, i = 1, 2, \ldots m\}$.

Now, we derive estimates and with the aid of Hörmander's results in [7], show that there exist constants A & B such that $h_1, h_2, \ldots, h_m$ is a set of generators of the algebra $A\varphi_r$ of holomorphic functions F on $\Omega\backslash Z_r$, such that if $x \in \Omega\backslash Z_r$

(*) $$|F(x)| \leqslant A \exp |B\varphi(x)|$$

Therefore there exist holomorphic functions $g_1, g_2, \ldots, g_m \in A\varphi_r$ in $\Omega\backslash Z_r$ with $\sum_{i=1}^{m} g_i h_i = 1$.

c) Next we use a lemma of Fornaess-Narasimhan [9], to show that each function $G_i := f_r^{\tau_r - 1} g_i$, $i = 1, 2, \ldots m$ extends holomorphically to all of Ω, where τ_r is the sheet number of

$$\Phi_r: \Omega \longrightarrow \mathbb{C}^n.$$

Since $f_r(p) \neq 0$ and $\sum_{i=1}^{m} g_i h_i = 1$, at least one of the sequences $G_i(z_n)$ diverges to ∞ as $h \longrightarrow \infty$, for $z_n \longrightarrow p$.

II. By I it is possible to construct a holomorphic function in Ω which cannot be extended holomorphically to any domain $D' \not\subset \Omega$, $D' \cap \Omega \neq \emptyset$. Theorem 1 follows from this.▲

Let f be holomorphic and nonconstant on any positive dimensional irreducible component of X. Then we have the following exact sequences of coherent analytic sheaves on X

$$1)\ 0 \longrightarrow f\mathcal{O} \xrightarrow{\ i\ } \mathcal{O} \xrightarrow{\ \pi\ } \mathcal{O}/f\mathcal{O} \longrightarrow 0$$

$$2)\ 0 \longrightarrow k \longrightarrow \mathcal{O} \xrightarrow{\ \sigma\ } f\mathcal{O} \longrightarrow 0$$

where σ is multiplication by f and k is the kernel of σ.

Theorem 2. *If $\Omega \subset X$ is open, and X is a Stein space or $\Omega \subset\subset X$, then Ω is Stein if and only if*

a) Cousin I problem is solvable in Ω (i.e., $\Gamma(\Omega, m) \longrightarrow \Gamma(\Omega, m/\mathcal{O})$) and

b) If f is a nonconstant holomorphic function in any irreducible component of X of positive dimension, then $\Omega \cap \{x \in X: f(x) = 0\}$ is Stein.

Proof. It is easy to see that the conditions are necessary. The proof of the sufficiency of the conditions consists of two steps: (I) Let f be as above and $Z = \{x \in X: f(x) = 0\}$. Given any

$$S \in \Gamma(\Omega \cap Z, \mathcal{O}/f\mathcal{O}|Z),$$

in the usual way we show that it is possible to extend it holomorphically to Ω. (II) We prove that Ω is holomorphically convex.

I) Let $\{U_i\}$ be an open Stein covering of $\Omega \cap Z$, with $U_i \subset \Omega$ open, and let $U_0 = \Omega \setminus Z$. For each U_i there exists $\Phi_i \in \Gamma(U_i, \mathcal{O})$ such that $\Phi_i|_{U_i \cap Z} = S|_{U_i}$.

Define for each U_i, $F_i = \Phi_i/f$, and $F_0 \equiv 1$ in U_0. Consider now the exact sequence (2) of sheaves on X. The restriction of (2) in each $U_i \cap U_j$, $i, j \in I$, induces the cohomology sequence:

$$\longrightarrow \Gamma(U_i \cap U_j, \mathcal{O}) \xrightarrow{\sigma^*} \Gamma(U_i \cap U_j, f\mathcal{O}) \longrightarrow H^1(U_i \cap U_j, k) \longrightarrow \dots$$

Hence
$$\Gamma(U_i \cap U_j, f\mathcal{O}) = f\Gamma(U_i \cap U_j, \mathcal{O}).$$

Since
$$S|_{U_i \cap U_j} = \Phi_i|_{U_i \cap U_j \cap Z} = \Phi_j|_{U_i \cap U_j \cap Z},$$

we have
$$\Phi_i - \Phi_j|_{U_i \cap U_j} \in \Gamma(U_i \cap U_j, f\mathcal{O})$$

So this difference is divisible by f in $U_i \cap U_j$, thus $F_i - F_j$ is holomorphic in $U_i \cap U_j$. We note that $F_0 - F_i$ is also holomorphic in $U_0 \cap U_i$, for all $i \in I$ and therefore the collection $\{U_i, F_i, i \in (I \cup \{0\})$ defines a Cousin I data in Ω, so by hypothesis there exists a meromorphic function F in Ω such that $F - F_i$ is holomorphic on U_i.

Let $g_i = F - F_i$, $i \in (I \cup \{0\})$. Define now $S^* = fF$. By definition of g_i, S^* is holomorphic on U_i, and is equal to S on $S \cap U_i$.

II) Let $\{z_n\}$ be a sequence in Ω without accumulation points in Ω. If X is Stein and $\{z_n\}$ has no accumulation points in Ω, by the holomorphic convexity of X, there exists $g \in \Gamma(X, \mathcal{O})$ unbounded in $\{\ _n\}$. If $\{z_n\}$ has an accumulation point in X, we can suppose that $z_n \longrightarrow z \in (Z \cap \mathrm{Fr}\,\Omega)$.

Because X is an holomorphically separable finite dimensional complex space, it is possible to find $f_1, \dots, f_m \in \Gamma(X, \mathcal{O})$ which separates points in X and satisfies:

$$\{z\} = \{x \in X: f_i(x) = 0, \ i = 1 \dots m\}$$

(see [1], Chap. V).

Let $Z_j = \{x \in X: f_j(x) = 0\}$. If for some $j = 1 \dots m$, $Z_j \cap \Omega = \emptyset$, then $1/f_j$ is unbounded in $\{z_n\}$. Otherwise suppose that $Z_1 \cap \Omega \neq \emptyset$. There is no loss of generality if we take f_1 as the function f in the first part of the proof.

Let $h_2, \dots h_m$ be the images of $f_2 \dots f_m$ in $(\Omega \cap Z, \mathcal{O}/f_1\mathcal{O}_1|_{Z_1})$. By hypothesis this space is Stein. Then there exist $g_2 \dots g_m \in \Gamma(\Omega \cap Y, \mathcal{O}/f_1\mathcal{O})$ such that $g_2 h_2 + \dots + g_m h_m = 1$. By Part (I) there exist $G_2 \dots G_m \in \Gamma(\Omega, \mathcal{O})$ induced by the $g_2 \dots g_m$ in $(\Omega \cap Y, \mathcal{O}/f_1\mathcal{O}_1|_{Z_1})$. Then we have

$$1 - G_2 f_2 - \dots - G_m f_m \in \Gamma(\Omega, f_1\mathcal{O}).$$

Observe that k is in a natural way an $\mathcal{O}/f_1\mathcal{O}|_{Z_1}$-module. By the Theorem B of Cartan

$$H^q(\Omega\cap Z_1, k|_{\Omega\cap Y}) \cong H^q(\Omega, k) = 0, \qquad q\geqslant 1$$

therefore, if we take the restriction of (2) to Ω, we obtain in cohomology $\Gamma(\Omega, f_1\mathcal{O}) = f_1\Gamma(\Omega, \mathcal{O})$. Hence there exists a function $G_1 \in \Gamma(\Omega, \mathcal{O})$ such that

$$1 = G_1 f_1 + G_2 f_2 + \ldots + G_m f_m .$$

Since all $h_1, \ldots h_m$ vanish on Z_1, at least one $G_1, \ldots, G_m$ is unbounded on the sequence $\{z_n\}$. Thus the sufficiency of the conditions is proved. ▲

Theorem 3. *Let Y_1, Y_2, be open Stein subsets of X, and let $Y = Y_1 \cup Y_2$*

a) If $\dim_{\mathbb{C}} H^1(Y, \mathcal{O}) < \infty$ and X is Stein on $Y \subset\subset X$, then Y is Stein.

b) If $H^1(Y, \mathcal{O}) = 0$ and X is irreducible, then Y is a domain of holomorphy.

Proof.

a) The proof consists of two steps: (I) we show easily that for any coherent analytic sheaf $\mathcal{S}$ on X, $H^q(Y, \mathcal{S}) = 0$, $q\geqslant 2$; (II) by induction on the dimension of X, we prove that Y is holomorphically convex.

I) Consider the Mayer-Vietoris sequence

$$\longrightarrow H^{q-1}(Y_1\cap Y_2, \mathcal{S}) \longrightarrow H^q(Y, \mathcal{S}) \longrightarrow H^q(Y_1, \mathcal{S}) \oplus H^q(Y_2, \mathcal{S}) \longrightarrow \ldots$$

By the theorem B of Cartan, it follows that $H^q(Y, \mathcal{S}) = 0$, $q \geqslant 2$.

II) If $\dim_{\mathbb{C}} X = 1$, the result follows immediately, since Y is noncompact. Suppose now the result is true for $n \leqslant k$ and let $\dim_{\mathbb{C}} X = k+1$. Let f be a nonconstant holomorphic function on any positive-dimensional irreducible component of X, such that $Z = \{x \in X:\ f(x) = 0\}$ has dimension k. Also suppose that $Y \cap Z \neq \emptyset$.

Consider now the exact sequence (1) over Y.

Since $H^2(Y, f\mathcal{O}) = 0$, the induced map $\Pi^*: H^1(Y, \mathcal{O}) \longrightarrow H^1(Y, \mathcal{O}/f\mathcal{O})$ is surjective. It follows that $\dim_{\mathbb{C}} H^1(Y, \mathcal{O}/f\mathcal{O}) < \infty$.

Note that
$$H^1(Y, \mathcal{O}/f\mathcal{O}) \cong H^1(Y\cap Z, \mathcal{O}/f\mathcal{O}|_Z)$$

and
$$Y\cap Z = (Y_1\cap Z)\cup(Y_2\cap Z)$$

From the induction hypothesis it follows that $Y\cap Z$ is holomorphically convex; therefore

$$H^q(Y\cap Z, \mathcal{O}/f\mathcal{O}|_Z) = 0, \qquad q\geqslant 1,$$

by Theorem B of Cartan.

Consider further the exact sequence (2) over Y. By Part (I), we have $H^2(Y, k) = 0$ and so $\dim_{\mathbb{C}} H^1(Y, fO) \leqslant \dim_{\mathbb{C}} H^1(Y, O) < \infty$.

From the exact cohomology sequence induced by (1) we have

$$\longrightarrow \Gamma(Y, O) \longrightarrow \Gamma(Y, O/_{fO}) \longrightarrow H^1(Y, fO) \longrightarrow H^1(Y, O) \longrightarrow H^1(Y, O/_{fO}) \longrightarrow$$

It follows that $\dim_{\mathbb{C}} H^1(Y, O) \leqslant \dim_{\mathbb{C}} H^1(Y, fO)$. Thus we have that i^* is a surjection from one finite-dimensional vector space to another of the same dimension and therefore i^* is an isomorphism. We conclude that given any section $S \in \Gamma(Y \cap Z, O/_{fO}|_Z)$, there exists $S \in \Gamma(Y, O)$ such that $S|_Z = S$. Now by an argument similar to that used in Step II of Theorem 2, we obtain the holomorphic convexity of Y.

b) As in Step I of (a), we have $H^q(Y, O) = 0$, $q \geqslant 2$. If Y is not a domain of holomorphy, by definition, there are domains $G, G' \subset X$ such that $G \subset G' \cap Y$, $G' \not\subset Y$, and for any $f \in \Gamma(Y, O)$ there exists $f' \in \Gamma(G', O)$ with $f|_G = f'|_G$.

Let $z_0 \in (F_r Y \cap G')$. Since f' is uniquely determined by f, we may define the map:

$$\alpha\colon \Gamma(Y, O) \longrightarrow \mathbb{C}, \quad \text{where} \quad \alpha(f) = f'(z_0)$$

is an algebra homomorphism.

Because X is a holomorphically separable finite dimensional complex space, there exist holomorphic functions $g_1, \ldots, g_m$ on X which separate points on X.

Consider now the following theorem of Nagel [2].

Theorem. *(Nagel). Let X be a topological space, and let S be a sheaf of local $\mathbb{C}$-algebras over X such that:*

a) For all $x \in X$, the composition $\mathbb{C} \longrightarrow S_x \longrightarrow S_x/m_x$ is an isomorphism, where m_x is the maximal ideal of S_x.

b) For every $f \in \Gamma(Z, S)$, the associated complex valued function $\hat{f}$ is continuous, where $\hat{f}$ is the residual class of the germ of f at x in S_x/m_x.

c) For all $q \geqslant 1$, $H^q(X, S) = 0$

Assume also that there exist a finite number of global sections $f_1, \ldots, f_m \subset \Gamma(X, S)$ such that the associated complex valued functions separate the points of X.

Then for every nonzero algebra homomorphism

$$\varphi\colon \Gamma(X, S) \longrightarrow \mathbb{C}$$

there exists $x \in X$ such that $\varphi(f) = \hat{f}(x)$ for all f in $\Gamma(X, S)$.

We choose $g \in \Gamma(X, O)$ so that $g(z_0) \neq g(w)$ where w is the point associated by the theorem of

Nagel to the homomorphism α defined above. Then it follows

$$g(w) = \hat{g}(w) = \alpha(g) = g(z_0)$$

which is a contradiction.▲

Theorem 4 *Let $\Omega_1 \subset \Omega_2 \subset \ldots$ be a sequence of open Stein subsets of* X, $\Omega = \bigcup_{j=1}^{\infty} \Omega_j$ *and* $\dim_{\mathbb{C}} H^1(\Omega, O) < \infty$. *Then Ω is Stein.*

Proof. By Grauert's Reduction Theorem, it is enough if we give the proof in the nonreduced case. We apply induction on the dimension of X.

If $\dim_{\mathbb{C}} X = 1$, the result follows immediately. Suppose now the result is true for $n \leqslant k$ and $\dim_{\mathbb{C}} X = k+1$. By a result of Andreotti-Visentini [3], if $q \geqslant 1$ and for every analytic coherent sheaf S over Ω we have $H^q(\Omega_j, S) = H^{q+1}(\Omega_j, S) = 0$, for all $j \geqslant 1$, then $H^{q+1}(\Omega, S) = 0$. This implies $H^q(\Omega, S) = 0$ if $q \geqslant 2$. We will also show that $H^1(\Omega, O) = 0$.

As before, let f be a nonconstant holomorphic function on any positive-dimensional irreducible component of X such that $Z = \{x \in X: f(x) = 0\}$ has dimension k, suppose that $Z \cap \Omega \neq \emptyset$.

Consider the exact sequence (1) over X

$$0 \longrightarrow fO \xrightarrow{\ i\ } O \longrightarrow O/fO \longrightarrow 0$$

Since fO and O are coherent sheaves on Ω, $\dim_{\mathbb{C}} H^1(\Omega \cap Z, O/fO|_Z) < \infty$. Observe that the sequence $\Omega_j \cap Z$ is an increasing exhaustion of k-dimensional open **Stein** subsets of $\Omega \cap Z$.

From the induction hypothesis, it follows that $(\Omega \cap Z, O/fO|_Z)$ is Stein, therefore $H^q(\Omega \cap Z, O/fO|_Z) = 0$, $q \geqslant 1$.

Consider now the exact sequence(2) over Ω.

Then we obtain by composition of i with σ in the sequences (1) and (2) a surjective homomorphism γ_f: $H^1(\Omega, O) \longrightarrow H^1(\Omega, O)$. But $\dim_{\mathbb{C}} H^1(\Omega, O) < \infty$ and hence γ_f^* is an isomorphism.

Set $a \in H^1(\Omega, O)$ and suppose we have a relation $\sum_{m=1}^{k} C_m \gamma_{f^m}^*(a) = 0$. It follows that there exists a polynomial P(f) such that $\gamma_{P(f)}^*(a) = 0$. But applying the same arguments to P(f) in place of f, we conclude that $\gamma_{P(f)}^*$ is also an isomorphism. Since P(f) cannot be constant in any positive-dimensional irreducible component of X, $a = 0$, i.e., $H^1(\Omega, O) = 0$.

We can now prove the holomorphic convexity of Ω directly in the case X Stein or $\Omega \subset\subset X$ by applying the same method of Step II of Theorem 2. For that we must observe that $H^1(\Omega, O) = 0$ implies $\Gamma(\Omega, O) \twoheadrightarrow \Gamma(\Omega, O/fO)$. Alternatively, it is possible to obtain the holomorphic convexity of Ω by applying a theorem of Markoe [4].▲

REFERENCES

[1]. R. C. Gunning-H. Rossi, *Analytic Functions of Several Complex Variables,* Prentice-Hall, N. J., (1965).

[2]. A. Nagel, *"Cohomology, Maximal Ideals and Point Evaluations",* **Proc. Amer. Math. Soc.** 42 (1974), 47-50.

[3]. A. Andreotti-E. Vesentini, *"Les théorèmes foundamentaux de la théorie des espaces holomorphiquement complets",* Seminaire Ehresmann", 4, Paris, (1962-63), 1-31.

[4]. A. Markoe, *"Runge Families and Inductive Limits of Stein Spaces",* **Ann. Inst. Fourier** 27, Fasc. 3 (1977).

[5]. A. Andreotti-R. Narasimhan, *"Oka's Heftungs Lemma and The Levi Problem for Complex Spaces",* **Trans. Amer. Math. Soc.** III, (1964), 354-366.

[6]. K. Oka, *"Domaines finis sans point critique interieur",* **Japan J. Math.** 23 (1953), 97-155.

[7]. Hörmander, *"Generators for Some Rings of Analytic Functions",* (**Bull. Amer. Math. Soc., Vol.** 73, (1967).

[8]. H. Skoda, *"Applications des techniques L^2 à la théorie des ideaux d'une algebra de fonctions holomorphes avec poids",* **Ann. Scient. Ec. Norm. Sup.** 4eme. Serie, E-5, (1972) 545-579.

[9]. J. E. Fornaess-R. Narasimhan, *"The Levi Problem on Complex Spaces with Singularities",* **Math. Ann.** 248, (1980), 47-72.

[10]. E. Ballico, *"Anullamento di gruppi di coomologia e spazi di Stein",* **Bulletino U.M.I.** (5), 18-B (1981), 649-662.

[11]. J. E. Fornaess, *"An Increasing Sequence of Stein Manifolds Whose Limit is Not Stein",* **Math. Ann.** 223 (1976), 275-277.

Centro de Investigación y de Estudios Avanzados del IPN
Mexico City, Mexico

Escuela Superior de Física y Matemáticas
Instituto Politécnico Nacional
Mexico City, Mexico